# GLOSSARY OF PLANT TISSUE CULTURE

# GLOSSARY OF PLANT TISSUE CULTURE

Danielle J. Donnelly
Department of Plant Science
McGill University
Ste. Anne de Bellevue
Quebec, Canada

William E. Vidaver
Department of Biological Sciences
Simon Fraser University
Burnaby
British Columbia, Canada

Advances in Plant Sciences Series
VOLUME 3
Theodore R. Dudley, Ph.D., General Editor

**DIOSCORIDES PRESS**
*Portland, Oregon*

ISBN 0-931146-12-7
Printed in Hong Kong

DIOSCORIDES PRESS
9999 SW Wilshire
Portland, Oregon 97225

**Library of Congress Cataloging-in-Publication Data**

Donnelly, Danielle J.
Glossary of plant tissue culture / Danielle J. Donnelly, William E. Vidaver.
p. cm. -- (Advances in plant sciences series ; v. 3)
Bibliography: p.
ISBN 0-931146-12-7
1. Plant tissue culture--Dictionaries. I. Vidaver, William E. II. Title. III. Series.
QK725.D64 1988
581.8'2'0321--dc19 88-9591
CIP

# Contents

# Dedication

This glossary is dedicated to Dr. Toshio Murashige.

# Preface

This glossary has been prepared to serve as a comprehensive guide for interpreting the current literature pertaining to plant cell and tissue culture. It is intended to supplement rather than supercede prior technical publications such as *Plant Propagation by Tissue Culture* (George and Sherrington, 1984) or *Handbook of Plant Cell Culture* (Evans et al., 1983) and to complement rather than replace the glossary in *In Vitro* 20:19–24 (Tissue Culture Association Terminology Committee, 1984). The terms included in the glossary have been selected by examination of textbooks, journals and glossaries dealing entirely or in part with plant tissue culture or related fields. References consulted during preparation of the manuscript are listed under "Sources" on pp 139–141.

Each entry is listed alphabetically, in **bold faced type**, at the point in the text where it is defined. The more common derivatives of a term are also given in **bold faced type**. Multi-sense entries are separated by numbers (**1, 2**, etc.). Some common English terms which have particular meanings in plant tissue culture, such as "sterilize" are defined in the text. Commonly used chemicals are listed alphabetically by their chemical names followed by their abbreviation, chemical formulae and atomic (aw) or molecular (mw) weights.

A publication of this nature should be periodically updated and revised; the authors will welcome all suggestions for amendments or additions.

We wish to thank Ms. Cindy Smith for typing and proofreading the manuscript, as well as Mrs. Lydiane Kyte and Dr. William Pengally for reviewing the final manuscript for technical accuracy. Ms. Pauline Kaye, Mr. Eric Cabot, Ms. Rebecca Dalby and Mr. Wayne Donnelly are also thanked for their contributions.

# A

**ABA:** See **abscisic acid**.

**abaxial:** The side or on the side distant from the axis; as on the lower surface of a leaf.

**aberrant:** A growth that deviates from the normal or usual type; as exceptional growths (**aberrations**) which occur in some tissue cultures.

**abort:** To stop or fail to reach full development (maturity) or to develop imperfectly; as some hybrids are prone to embryo **abortion** or as some attempts to culture a plant may fail (**abortive** attempt).

**abscisic acid (ABA, abscisin, dormin, [S-(Z,E)]-5(1-hydroxy-2,6,6-trimethyl-4-oxo-2-cyclohexen-1-yl)-3-methyl-2,4-pentadienoic acid, $C_{15}H_{20}O_4$ mw 264.31):** A naturally occurring plant hormone or growth regulator (sesquiterpenoid) displaying growth-inhibiting action. The relative balance between ABA and the growth promoters (auxins, gibberellins and cytokinins) may determine the response. Its mode of action involves the inhibition of protein and nucleic acid synthesis. It is associated with apical dominance, the onset and maintenance of dormancy, promotion of flowering, as well as promotion of senescence, abscission of leaves and fruit and stomatal closure under stress (especially water stress) situations.

**absciss:** The dropping of leaves, flowers, fruits or other plant parts, usually following the formation of an **abcission** zone or separation layer which has a role in the separation (abscission) of that plant organ from the plant body. The balance between auxins and abscisic acid may control abscision. Gibberellins have also been implicated.

**absolute alcohol:** 99% pure ethanol (by weight). See **ethanol**.

**absolute requirement:** A nutritional or environmental substance or condition that is necessary for plant growth and development.

**absorb:** **1.** To drink up or suck, imbibe. The capacity for **absorption** (**absorptive** capacity) is inherent in some materials (**absorbents**) such as cotton wool (**absorbent** cotton). **2.** To take in and utilize; as plants absorb water and nutrients through root hairs, or light, utilizing it in photosynthesis. In active absorption solute is taken up against an osmotic gradient, using energy. In passive absorption energy is not required; as in water intake, which is controlled by the transpiration rate.

**acaricide:** A pesticide used to kill or control mites or ticks. Mites occasionally infest culture rooms and cultures, spreading pathogens. **Acaricidal** paint is applied to shelves and walls. Infected or old cultures must be discarded for mite control.

**accessory bud:** A bud located above or beside the main axillary bud.

**acclimate or acclimatize:** To adapt or adjust physiologically to a new environment or climate. The process is **acclimation** or **acclimatization** and is regulated by natural processes or human cultural practices. Usage of acclimate may be limited to adaptive responses involving changes in complex environmental features, such as seasonal changes. Usage of acclimatize may be limited to adaptive responses involving changes in a selected environmental feature, such as light intensity.

**acetocarmine:** A dye which stains chromosomal material; used in chromosome analysis.

**acid:** A water soluble chemical compound that dissociates releasing hydrogen ions but not hydroxyl ions, which reacts with base (neutralizes) to form a salt and water; reddens litmus paper; and has a pH of less than seven. It is used to dissolve cytokinins and to adjust the pH of tissue culture media. **Acidic** substances have the properties of acids.

**acquired:** Developed in response to the environment, not inherited; as a character trait (acquired characteristic) resulting from environmental effect(s).

**acropetal:** **1.** Developing or blooming in succession towards the apex; as leaves or flowers develop **acropetally**. **2.** Refers to the transport or movement of substances towards the apex; as the movement of water through the plant. The opposite tendency is termed basipetal.

**activated charcoal or activated carbon:** Charcoal which has been treated to remove hydrocarbons and to increase its adsorptive properties is considered activated. It acts by condensing and holding a gas or solute onto its surface; as inhibitory substances in nutrient medium may be adsorbed to charcoal included in the medium. In addition, rooting factors such as phenolamines present as contaminants in charcoal may stimulate growth in vitro. Its addition to rooting medium may stimulate root initiation in some plant species. Activated charcoal may differ in origin and in composition. It is considered an undefined constituent of nutrient media.

**active absorption:** Energy-requiring uptake (absorption) of materials; such as minerals into a plant.

**adapt:** Adjust, or to become adjusted (structurally and/or functionally) to a new circumstance or environment; as explants must adapt to the culture environment. The process is **adaptation**. It has both evolutionary and physiological implications.

**adaxial:** The side or on the side nearest to the axis; as the upper surface of a leaf.

**addendum, pl. addenda:** A supplement, thing or substance to be added; a constituent; as complex addenda such as coconut milk or orange juice are included in some nutrient media.

**adenine ($C_5H_5N_5$ mw 135.14):** A white crystalline purine base present in DNA, RNA and nucleotides like ADP and ATP. A B group vitamin ($B_4$) generally available as **$C_5H_5N_5.3H_2O$ mw 189.13**. It is added to some tissue culture media, as **adenine sulphate**, to promote shoot formation and for its weak cytokinin effect. It may reinforce the effects of other cytokinins. It is present in plant tissues combined with niacinamide, phosphoric acids and D-ribose.

**adenine sulphate ($(C_5H_5N_5)_2.H_2SO_4.2H_2O$ mw 404.37):** A growth factor used in some plant tissue culture media. See **adenine**.

**adenosine ($C_{10}H_{13}N_5O_4$ mw 267.24):** A widely distributed mononucleoside; adenine plus a pentose sugar (D-ribose). **Deoxyadenosine** is adenine linked to 2-deoxy-D-ribose. It is isolated from yeast nucleic acid and is added to some tissue culture media for its weak cytokinin effect. It may also reinforce the effects of other cytokinins.

**adenosine diphosphate (ADP, $C_{10}H_{15}N_5O_{10}P_2$ mw 427.31):** A nucleotide of adenine, D-ribose and two phosphate groups. It functions as an important coenzyme in many biological reactions. It undergoes oxidative phosphorylation to adenosine triphosphate (ATP) in the mitochondria.

**adenosine monophosphate (AMP, $C_{10}H_{14}N_5O_7P$ mw 347.18):** A mononucleotide of adenine and D-ribose phosphate. One or two phosphoric acid groups can be esterified onto the phosphate to yield ADP and ATP.

**adenosine triphosphate (ATP, $C_{10}H_{16}N_5O_{12}P_3$ mw 491.19):** A ribonucleoside 5′-triphosphate. It functions as a donor of a phosphate group in living cells, with associated release of energy as the high energy phosphate bond is broken. ATP minus one phosphate group is adenosine diphosphate (ADP).

**adsorb:** The adhesion of a liquid, gas or dissolved material to a solid surface, resulting in concentration of the **adsorbate** (the **adsorbed** material) to the **adsorbent** (the **adsorbing** agent, such as activated charcoal). The process is **adsorption**.

**adult:** **1.** The reproductively mature phase of an organism or its tissues. The development of reproductive competence should not be confused with aging, as relatively young plants may be reproductively mature. The opposite of the juvenile phase. **2.** An individual organism in this phase.

**advanced:** Applied to an organism or a part thereof it implies considerable development from the ancestral stage or from the explant stage.

**adventitious:** **1.** Produced in an abnormal or unusual position, or at an unusual time of development or away from the natural habitat; as when plant organs (buds, shoots, roots, others) develop on callus or nonzygotic embryos (embryoids) develop without an ovary or fertilization on callus (**adventive** embryony). **2.** A manner of growth relying on adventive processes.

**adventive:** Formed or growing in an **adventitious** manner; as adventive embryony (nonzygotic embryo development) occurring in *Citrus* spp.

**aerate:** To supply with or mix with air or gas. The process is **aeration**.

**aerenchyma:** Tissues (parenchyma) comprised of thin-walled cells with large intercellular air spaces. Such cells are characterized by great mechanical strength utilizing a minimum amount of tissue. This is a common feature of some aquatic plants, to which this lends buoyancy.

**aerial roots:** Roots emerging from above-ground portions of the plant; as from the stem.

**aerobe:** An organism growing or occurring only in the presence of and utilizing free oxygen; as do green (**aerobic**) plants.

**aerobic or aerobiotic:** Refers to an organism that lives in or a process occurring in the presence of molecular oxygen.

**agamic or agamous reproduction:** Asexual reproduction without germ cell union.

**agar or agar-agar:** A gelatinous polysaccharide obtained from the red alga *Gelidium corneum* and from several other red algae. It is a solidifying agent that mixed with nutrient media (0.6–1%), forms a gel for growing tissue cultured plants and for other purposes. It ranges in quality from relatively inert to very impure (complex, undefined). Its firmness as a gelling agent is affected by medium pH (it is softer when the medium is more acidic) and salt concentration (it is softer when the medium is more dilute). Agar gels melt at about 100°C but solidify at about 44°C.

**age:** **1.** The period in the life cycle of an organism; the process of growing older, mature. **Aging** should not be confused with the development of reproductive maturity as relatively young plants may be reproductively competent. The reverse is also true. **2.** The state of being old or senescent. **3.** Culture age is a function of the number of subcultures and the time after subculture.

**agenesis:** The absence of development.

**agglutinate:** **1.** To cause to unite, adhere; as if glued. **2.** To gather into a clump or mass; as protoplasts and bacteria, in the presence of specific antibodies, tend to stick to one another. The process is **agglutination**.

**agglutinin:** An antibody capable of clumping bacteria or other cells.

**aggregate:** **1.** A clump or mass formed by gathering or collecting units. **2.** A body of loosely associated units or parts; as a friable callus or cell suspension is a loose association of cells. **3.** A coarse inert material such as gravel that is mixed with soil to increase its porosity. **4.** A serological reaction in which the antibody and antigen react and precipitate out of solution.

**agitate:** To move a solution with an irregular rapid motion; as the contents of fermentors or flasks on shakers in order to maximize tissue or cell exposure to nutrients, facilitate gaseous exchange (aeration) and to disperse cells.

**agriculture:** Both a science and an art of cultivating the soil, tilling, farming, raising food, livestock, etc. The occupation of **agriculturalists**; farmers, experts in agriculture.

***Agrobacterium tumefaciens*:** The bacterium causing crown gall disease of plants, inducing tumors to form. Tumors may be cultured in vitro and are useful in the study of this disease. The Ti plasmid of *A. tumefaciens* is known to cause the disease and a small portion of it is used as a vector in the genetic modification of higher plant cells. Novel DNA sequences are spliced into the plasmid DNA segment, the DNA is circularized and introduced into the cultured plant cells.

**agronomy:** The science of land cultivation. The occupation of **agronomists**.

**alanine (Ala, $C_3NO_2H_7$ mw 89.10):** One of the 20 common amino acids found in proteins. Occasionally added to plant tissue culture media.

**albido:** The white tissue beneath the peel of citrus fruits.

**albino:** An organism lacking normal pigmentation, due to genetic factors; as in the white portion of variegated leaves. The condition is **albinism**. A conspicuous plastome (plastid) mutant involving loss of chlorophyll; sometimes a product of plant tissue culture.

**alcohol:** A colorless, inflamable, organic (liquid compound characterized by the presence of a functional hydroxyl (OH) group, as in ethanol. **Alcohols** are used as solvents, fixatives, disinfecting agents and for many other purposes.

**alcohol, anhydrous:** See **ethanol** or **ethyl alcohol**.

**alcohol, denatured:** Adulterated ethanol used for a variety of industrial purposes and unfit for drinking purposes owing to the addition of any of: methanol, camphor, benzene, acetone, sulphuric acid or one of many other substances.

**aleuroplast:** A colorless plastid (leucoplast) involved in protein storage and found in many seeds.

**aliquot or aliquot part:** An evenly divided unit, portion or sample (fraction) of the whole.

**alkaline:** A basic solution with a pH above 7.0. See **base**.

**alkaloid:** **1.** An organic compound with alkaline properties (usually poisonous), produced by some plants and containing carbon, hydrogen and nitrogen (and most often, oxygen). **2.** The active component of many plant-derived drugs and poisons. These include nicotine, quinine, cocaine, morphine and others. Alkaloid production is a common objective of cell culture and secondary product synthesis studies. Alkaloid function in plants remains conjectural; they may be by-products of metabolism, or may provide protection against animals that feed on plants.

**allele or allelomorph:** One of two or more alternative states of a gene differentially affecting developmental processes. These are **allelic** or **allelomorphic**; occupying the same locus (position) on homologous chromosomes, and separated from one another at meiosis. When **alleles** are present in pairs one is often dominant to the other one (recessive). The wild-type allele is usually dominant. Recessives may result from mutation and are usually deleterious. The term **allelomorph** is used for the trait produced by an allele.

**allograft:** See **homograft**.

**allopolyploid or alloploid:** A polyploid with one or more chromosome sets from different genera, species or strains.

**aluminum or aluminium (Al aw 26.9815 an 13):** An abundant light, white, maleable and ductile metal. It occurs naturally as **aluminum silicate** or **aluminum oxide**. It is added to some tissue culture media as **aluminum chloride** or **aluminum sulfate**. However, the role of aluminum in microelement formulations has not been adequately demonstrated.

**ambient:** The environment at a particular time; as that set of climatic conditions existing during an experiment.

**amide:** An organic compound formed when hydrogen atoms of ammonia ($NH_3$) are replaced by acyl residues, as in acetamide ($CH_3CONH_2$). The general formula is $RCONH_2$, and $-CONH_2$ is the amide group.

**amino acid:** One of many organic acids containing a basic amino group ($NH_2$), and an acidic carboxyl group (COOH). Of several hundred naturally occurring **amino acids** only 20 are commonly found in protein molecules, linked by peptide bonds. Some are commonly added to plant tissue culture media, especially glycine and glutamine.

**aminoacetic acid:** See **glycine**.

**3-(2-aminoethyl)indole:** See **tryptamine**.

**aminopurine:** A purine nitrogeneous base subunit of nucleotides and nucleic acids having an amine (-$NH_2$) group ($N^6$) on the number 6 carbon atom.

**4-amino-3,5,6-trichloro-2-pyridinecarboxylic acid:** See **picloram**.

**amitosis:** Cell division (cytokinesis), including nuclear division through constriction of the nucleus, without chromosome differentiation as in mitosis. The maintenance of genetic integrity and diploidy during amitosis is uncertain. This occurs in the endosperm of flowering plants.

**ammonium ion:** $NH_4^+$.

**AMP:** See **adenosine monophosphate**.

**amphidiploid or amphiploid:** An interspecific hybrid with one of each kind of parental chromosomes per cell. Usually sterile.

**amphistomatic:** Leaves having stomata on both surfaces.

**ampoule or ampule or ampul:** A small sealed glass vessel or any similar vial containing a solution for a single dose, usually from a hypodermic syringe.

**amylase:** One of a class of enzymes that hydrolize starch into disaccharides and glucose.

**amyloplast or amyloplastid:** A starch synthesizing and storing plastid (leucoplast) found in plant storage organs; as in the endosperm of seeds.

**amylose:** See **starch**.

**anaerobe:** An **anaerobic** organism; as one that lives and grows in the absence of free oxygen.

**anaerobic or anaerobiotic:** **1.** The lack of oxygen. **2.** May refer to organisms or processes not requiring oxygen or proceeding in its absence.

**analog or analogue:** **1.** Something similar, resembling or comparable (**analogous**) to something else. **2.** An organ similar in function, structure or appearance but different in origin or ancestry to that of another plant. **3.** A compound similar in chemical structure and in effect to another; as the synthetic growth regulators are to the natural plant hormones.

**anaphase:** The third meiotic or mitotic phase, during which the chromatids (halves of the sister chromosome pairs) or homologous chromosomes move from the metaphase plate (equatorial plane) to their respective (opposite) poles of the spindle.

**anatomy:** Plant structure or the study of detailed internal plant structure as; that of cells, tissues and organs.

**androecium:** The collective term for the stamens of a flower.

**androgenesis:** Plant development from male gametophytes. **1.** Plant (haploid) development from pollen in pollen cultures. **2.** The development of haploid individuals from sperm following disintegration of the egg; as in some tobacco varieties.

**anergized culture:** See **habituated culture**.

**aneuploid:** Individual cell or organism in which genetic change has occurred resulting from addition or subtraction of individual chromosomes and occasionally compounded by chromosomal rearrangements. The condition is **aneuploidy** or heteroploidy. Monosomics, trisomics and tetrasomics are aneuploid.

**aneurine:** See **thiamine**.

**aneurine hydrochloride:** See **thiamine hydrochloride**.

**angiosperm:** A member of the group of flowering vascular plants (**Angiospermae**) whose seeds are enclosed within a mature ovary (fruit) in contrast to the seeds of gymnosperms which are not enclosed in an ovary. It contains about 250,000 species in two groups, the Monocotyledonae and Dicotyledonae, with one and two cotyledons in the embryo respectively.

**anhydrous:** Without water, dry.

**anion:** A negatively-charged ion in solution.

**anomalous:** An exception to the rule, type or form; as irregular or abnormal growth or performance in cultured plants as compared to controls.

**anther:** The upper part of a stamen that contains the pollen grains in pollen sacs (anther sacs). A common explant source for plant tissue culture (anther culture) aimed at the production of monoploid plants.

**anther culture:** Refers to culture of single pollen grains or of the anther, containing the male gametophytes or microspores, with the objective of producing monoploid plants.

**anthesis:** The flowering period or efflorescence. This is the time of full bloom which lasts till fruit set.

**anthocyanin:** One of a group of water soluble pigments, of soluble glycosides, varying in color, present in plant cell sap and common in flowers.

**antiauxin:** A chemical that interferes with the auxin response. These may or may not involve prevention of **auxin** transport or movement in plants. Some are said to promote morphogenesis in vitro; as 2,3,5-triiodobenzoate (TIBA mw 499.81), or 2,4,5-trichlorophenoxyacetate (2,4,5-T mw 255.49), which stimulate the growth of some cultures.

**antibiotic:** One of many natural organic substances (or their synthetic analogues) secreted by plants or microorganisms that are toxic to other species, retard or prevent their growth and presumably function as defense mechanisms; as bacitracin, gentamycin, mycostatin, nystatin, penicillin, phosphomycin, rifampicin, streptomycin and terramycin etc. The phenomenon is **antibiosis**. **Antibiotics** are sometimes included in plant tissue culture media, with varying results, ranging from dramatic culture stimulation to induction of chromosomal instability.

**antibody:** A highly specific protein produced in the blood of a mammal in response to an injected foreign antigen. **Antibodies** can be extracted from the blood and used for immunological assays (immunoassays).

**anticlinal:** The plane of cell division or the cell wall oriented perpendicular (at right angles) to the surface of an organ.

**anticodon:** See **codon**.

**antigen:** A substance inducing protein (antibody) formation with which the protein or carbohydrate substance (antigen) reacts specifically; as purified plant viruses are injected into rabbits to induce specific antibody formation which can then be extracted from rabbit blood and used in plant virus assay work (serological assays).

**antioxidant:** A substance (such as ascorbic acid, citric acid or others) which is sometimes added to the sterilizing solution or to isolation medium to inhibit or prevent oxidative browning of the culture medium, due to bleeding of phenolic exudates. The latter may lead to tissue necrosis and death. Ascorbic acid (100 mg/1) and citric acid (150 mg/1) are most commonly used in sterilizing solutions.

**antipodal:** Opposite the micropylar end of a developing embryo sac are three haploid nuclei. The function of these **antipodals** is unknown and at fertilization they may either multiply and enlarge or disintegrate.

**antiseptic:** The process or agent capable of inhibiting the growth of microorganisms.

**antiserum:** A blood serum of an animal containing antibodies (to some specific antigen) and providing immunity from a specific disease.

**apex:** The most extreme point of growth of a plant; as the **apical** shoot and root tips are located at the **apexes (apices)**, and contain the apical meristem.

**apical:** Located at the apex. The apical shoot tip is a common explant for plant tissue culture. This term also refers to the root apex, a less common explant.

**apical cell:** A meristematic initial in the apical meristem of shoots or roots of plants. As this cell divides new tissues are formed.

**apical dominance:** The terminal bud influence, a widespread phenomenon in the plant kingdom, which is related to the auxin content of apical buds, and is exerted on lateral buds, resulting in their growth suppression. Abscisic acid may also inhibit lateral bud growth. Cytokinins tend to promote lateral bud development, thereby overcoming apical dominance.

**apical meristem:** The meristematic cells at the stem or root growing tip and their recent derivatives. This is the region of primary tissue initiation either vegetative or reproductive; as the floral meristem. A common explant for plant tissue culture is the apical shoot tip, or the apical meristem tip. The meristem (apical or lateral) which implies the meristematic dome with no adjacent leaf primordia or stem tissue is rarely employed as an explant, except for virus elimination purposes, and even then the meristem tip is the usual explant.

**arabinose ($C_5H_{10}O_5$ mw 150.13):** A pentose sugar. An occasional carbohydrate additive in plant tissue culture media.

**arginine (Arg, $C_6H_{14}N_4O_2$ mw 174.20):** An amino acid important in histone proteins and occasionally added to plant tissue culture media, as **arginine hydrochloride**.

**arginine hydrochloride ($C_6H_{14}N_4O_2.HCl$ mw 210.67):** An amino acid salt used in some plant tissue culture media as a source of reduced nitrogen.

**artifact:** A product of extraneous, often human, agency that would not occur in nature; artificially induced in preparation for or during investigation, or the result of post-mortem changes.

**artificial seed:** Encapsulated or coated somatic embryos (embryoids) that are planted and treated like seed.

**artificial selection:** Plant selection by man for agronomic qualities.

**ascorbic acid or vitamin C ($C_6H_8O_6$ mw 176.12):** A water soluble vitamin present naturally in some plants and also synthetically produced. Aside from its use as a vitamin, it is used as an antioxidant in plant tissue culture; included in disinfection solutions and sometimes in media.

**aseptic:** Free of pathogens, contaminants, algae, bacteria, fungi, viruses, etc.; absence of all microorganisms. **Asepsis** is a fundamental requirement for plant tissue culture (aseptic culture).

**asexual:** Lacking or not involving **sex**; as in vegetative plant propagation.

**asexual propagation:** The multiplication of plants using a vegetative plant part or a portion thereof.

**asexual reproduction:** A reproductive process not involving union of gametes such as reproduction by spores not associated with sexual processes. Plant tissue culture propagation is for the most part carried out asexually (vegetative propagation or vegetative reproduction).

**asparagine (Asn, $C_4H_8N_2O_3$ mw 132.12):** An amino acid occasionally included in plant tissue culture media, as a source of reduced nitrogen.

**aspartic acid (Asp, $C_4H_7NO_4$ mw 132.12):** An amino acid necessary for nucleotide synthesis and occasionally included in plant tissue culture media.

**aspirate:** To draw something in or out, up or through using suction or a vacuum; as **aspiration** (vacuum) may be used in the disinfection process to draw disinfectant into the surface layers of plant tissue.

**assay:** **1.** To test or evaluate. **2.** The substance to be analyzed or the process of examining or testing it (chemically or by other means).

**atomic number (an):** The number of protons (positive charges) in an atomic nucleus or the number of electrons rotating around the nucleus of the neutral atom of an element.

**atomic weight (aw):** The relative atomic weight of an atom, compared to that of the common isotope of oxygen (aw 16.0).

**ATP:** See **adenosine triphosphate**.

**atrophy:** Reduced or diminished organ (or organism) size, shape or function. Usually a deteriorative change.

**atypical:** Refers to an appearance or function not conforming to type.

**autoclave: 1.** An enclosed chamber in which to heat substances under pressure to above their boiling points, to sterilize utensils, liquids, glassware, etc., using steam. The routine method uses steam pressure of $103.4 \times 10^3$ Pa at 121°C for 15 minutes, or longer for large volumes to reach temperature. **2.** A pressure cooker. These are employed in medium and instrument sterilization for plant tissue culture work. Over-sterilization degrades culture media constituents and caramelizes sugars, so is to be avoided. Under-sterilization results in culture media or equipment contamination. Heat labile components of tissue culture media cannot be **autoclaved**. These are commonly filter-sterilized. **3.** To carry out the process of **autoclaving**.

**autoploid:** A cell or individual with a characteristic number of chromosome sets, all of which are homologous. The condition is **autoploidy**.

**autopolyploid:** A cell or individual with in excess of two of the monoploid chromosome sets (all homologous) that characterize the species. Those with an odd number of chromosome sets are usually sterile. Those with an even number of chromosome sets may have reduced fertility. The condition is **autopolyploidy**.

**autotetraploid:** A cell or individual with four of the monoploid chromosome sets (all homologous) that characterize the species.

**autotroph or lithotroph:** An organism capable of manufacturing all of its own food (self sufficient) by building its own macromolecules from simple nutrient molecules ($CO_2$, inorganic nutrients); as do most green plants and some bacteria. The energy source for synthesis is either light, for photosynthesis (photoautotroph) or via chemosynthesis (chemoautotroph). Tissue cultured plants must grow **autotrophically** (phototrophically) once transferred from culture to soil (ex vitro).

**auxin:** One of a large class of plant hormones (phytohormones) allegedly produced in the growing tips of stems and roots. These, and many analogs to them, have been chemically synthesized. **Auxins** are implicated in apical dominance; the suppression of lateral bud development by the apical bud. They promote root initials. They promote growth through cell elongation (extension) rather than cell division. Auxins function in increasing cell wall plasticity, and so extensibility. They resemble 1*H*-indole-3-acetic acid (IAA) (the type member) in physiological activity so are used in plant tissue culture work to promote new cell division and enlargement, adventitious bud formation and rooting. The most commonly used in horticulture and research include IAA, 1*H*-indole-3-butanoic acid (IBA), 1-naphthaleneacetic acid (NAA) and (2,4-dichlorophenoxy)acetic acid (2,4-D). Stock solutions of these compounds are prepared by dissolving in a base,

KOH or NaOH (ca. 1 M), then making up to volume with water. Such stock solutions are usually refrigerated and kept in amber or other dark bottles as some of these hormones are relatively unstable and light sensitive.

**auxin-cytokinin ratio:** The relative proportion of auxin to cytokinin present in plant tissue culture medium. Varying the relative amounts of these two hormone groups in tissue culture formulas affects the proportional growth of shoots and roots in vitro. As the ratio is increased (increased auxin or decreased cytokinin contact), roots are more likely to be produced, and as it is decreased root growth declines and shoot initiation and growth are promoted. This relationship was first recognized by C.O. Miller and F. Skoog in the 1950's.

**auxotroph:** A microorganism or plant cell line possessing atypical nutritional requirements for some item(s) (growth factors) in addition to those required by the control or wild type which can synthesize them; as **auxotrophic** cells requiring certain amino acids.

**availability:** A reflection of the form and location of nutritional elements and their suitability for plant absorption. In plant tissue culture media this is related to the abundance of each nutritional element, the osmotic concentration and pH of the medium, the stability and solubility of the item in question, the presence of adsorbing agents in the media and other factors.

**axenic:** A pure culture of one species. This implies that cultures are free of microorganisms (aseptic or germ free).

**axil:** The angle formed by a lateral organ (leaf, branch, pedicel, etc.) and the upper (distal) side of the branch or stem. This is the site of **axillary** buds.

**axillary:** **1.** Pertaining to or situated in an axil; as a bud (axillary bud) or branch (axillary branch) occurring in the axil of a leaf. Axillary organs are derived from axillary or lateral meristems. **2.** Promotive of the growth of lateral buds; as are certain culture media and protocols.

**axillary bud proliferation:** Propagation in culture by protocol and media which promotes axillary (lateral shoot) growth. This is a technique for mass production (micropropagation) of plantlets in culture, achieved primarily through hormonal inhibition of apical dominance and stimulation of lateral branching.

**axis:** The main plant stem.

# B

**B:** The chemical symbol for the element **boron**.

**B vitamin:** One of a complex of vitamins made in plants and thought to be essential for healthy growth. Some are routinely added to plant tissue culture media to promote growth. The latter include thiamine **($B_1$)**, niacin **($B_3$)**, adenine **($B_4$)** and pyridoxine **($B_6$)**. Also part of the $B_6$ vitamin complex, although less often used in plant tissue culture media are pyridoxal 5-phosphate ($C_8H_{10}NO_6P$ mw 247.15), pyridoxal.HCl ($C_8H_9NO_3$.HCl mw 370.8) and pyridoxamine dihydrochloride ($C8H_{12}N_2O_2$.2HCl mw 241.12).

**B5:** See **Gamborg, O.L., R.A. Miller, and K. Ojima (1968)**.

**BA or BAP:** See *N*-(phenylmethyl)-1*H*-purin-6-amine or **6-benzylaminopurine**.

**backcross:** To breed a hybrid to one of its parents or to a genetically equivalent individual; as intergeneric somatic hybrids may be crossed with one or the other of the parental types, once the hybrids have been regenerated to plants, or (back fusion) may be accomplished at the protoplast level.

**bactericide:** A substance or agent that kills **bacteria** (usually rapidly); is **bactericidal**.

**bacteriostat:** A chemical or other agent that does not kill but prevents bacterial growth and multiplication; is **bacteriostatic**.

**bacterium, pl. bacteria:** Any of a large group of unicellular, prokaryotic, microscopic organisms in the division **Bacteria** or Prokaryota. In this group are some parasites, saprophytes and autotrophs. Included are some that are valued as fermenters. Bacteria may be photosynthetic but lack chlorophyll a and multiply by fission. In plant tissue culture bacteria are actively excluded through aseptic practices.

**bacticinerator:** An electrical appliance, capable of generating elevated temperatures, into which metal instruments are inserted and held while sterilization occurs. **Bacticinerators** are generally used to facilitate aseptic operations within laminar air flow cabinets, and may replace the bunsen burner, especially in areas without access to natural gas.

**balance:** **1.** The act of, or the instrument used for weighing (**balancing**). **2.** To adjust, equilibrate or proportion.

**Ball, E. (1946):** The first to obtain plants from cultured shoot apices.

**ballast:** A step-up transformer that increases the voltage and decreases the amperage to lighting systems such as fluorescent, high pressure sodium and metal halide lamps. **Ballasts** generate a substantial amount of heat so may be isolated from the culture area in commercial operations.

**BAP:** See **6-benzylaminopurine**.

**basal:** **1.** Located at the **base** of a plant or a plant organ. **2.** A fundamental medium formulation.

**basal nutrient medium:** See **medium formulation**.

**base: 1.** A water soluble chemical compound that may dissociate releasing hydroxyl ions, but not hydrogen ions; reacts with acid to form a salt; turns litmus paper blue; has a pH of more than seven; and is used to dissolve auxins and gibberellins and to adjust the pH of plant tissue culture media. **2.** The lowest part of an organism or a structure upon which it rests. **3.** A building block of nucleic acid molecules; purine or pyrimidine base.

**base pairing:** The pairing between complementary nitrogeneous bases in nucleic acid chains.

**basipetal: 1.** Developing or blooming successively towards the **base**; as leaves or flowers which develop **basipetally**. 2. Transporting or moving substances towards the base. The antonym is acropetal.

**batch culture:** A cell suspension grown in liquid medium of a set volume. Inocula of successive subcultures are of similar size and cultures contain about the same cell mass at the end of each passage. Cultures commonly exhibit five distinct phases per passage; a lag phase follows inoculation, then an exponential growth phase, a linear growth phase, next a deceleration phase and finally a stationary phase.

**bell or dispensing bell:** A glass apparatus for dispensing sterile distilled water used in tissue disinfection.

**benzalkonium chloride or zephiran chloride:** An alkyldimethylbenzylammonium chloride mixture sometimes employed as a disinfecting agent for plant material.

**6-benzylaminopurine or 6-benzyladenine (BAP or BA, *N*-(phenylmethyl)-1*H*-purin-6-amine, $C_{12}H_{11}N_5$ mw 225.20):** A synthetic cytokinin, and one of the most active, commonly employed in plant tissue culture media to induce axillary bud proliferation or callus proliferation.

**6-(benzylamino)-9-(2-tetrahydropyrany1)-9H-purine or *N*-(phenylmethyl)-9-(tetrahydro-2*H*-pyran-2-yl)-9*H*-purin-6-amine (PBA, $C_{17}H_{19}N_5O$ mw 309.40):** A synthetic cytokinin sometimes used in plant tissue culture media to induce axillary bud proliferation or callus proliferation.

**Bergmann, L. (1960):** Was first to obtain callus by plating cells from suspension cultures onto solid medium. This plating involved mixing cells with warm agar medium just prior to gelation in a petri dish (Bergmann plating technique).

**biennial:** A plant with a 2 year life cycle; growing vegetatively and storing food the first year and producing flowers and seeds the second year.

**binucleate:** Having two nuclei per cell.

**bioassay:** A biological assay or assessment procedure, performed on living cells or on a living organism; sometimes used to detect minute amounts of substances which influence or are essential to growth.

**biosynthesis:** Biological synthesis, the building or forming of biochemical compounds in a living organism.

**biotechnology:** The industrial use of biological processes; as yeast fermentation for alcohol production or plant cell culture for extraction of secondary products. The scope of biotechnology has been dramatically increased through genetic engineering.

**biotin ($C_{10}H_{16}O_3N_2S$ mw 244.31):** Part of the vitamin B complex. It is also called vitamin H. It is present in all living cells, bound to polypeptides or proteins, and is of import in fat, protein and carbohydrate metabolism. It is a common addition to plant tissue culture media.

**blade:** **1.** The lamina, or expanded part of a leaf or petal. **2.** The cutting end of a knife or scalpel.

**bleach:** A fluid, powder or other whitening (**bleaching**) or cleaning agent. Bleach contains calcium hypochlorite or sodium hypochlorite and is a common disinfectant used for cleaning working surfaces, tools and plant materials in plant tissue culture.

**bleeding:** **1.** Leaching of phenolic exudates, into the culture medium, that when oxidized turn black or purple. These exudates may foul the medium, either causing or coincident with culture necrosis and death. **2.** Loss of sap through wounding.

**Bn:** Abbreviation for backcross generation (first is **B1**, second is **B2**, etc.).

**boric acid or boracic acid or orthoboric acid ($H_3BO_3$ mw 61.84):** A microelement **boron** salt commonly used in plant tissue culture media. It also has mild antiseptic properties.

**boron (B aw 10.811 an 5):** A chemical element which is usually in the form of a brown amorphous powder or yellow crystal. It occurs as **borax** or **boric acid** and is a metalloid microelement implicated in calcium uptake. Deficiency symptoms involve internal stem browning. It is included in plant tissue culture media as **boric acid**.

**borosilicate:** Soda-lime glass that contains boric oxide (ca. 5%). It is very strong, and relatively high impact, abrasion and heat resistant as compared to soda-lime glass.

**botany:** The scientific study of plants.

**bract:** A leaflike structure subtending a flower or inflorescence.

**branch:** **1.** An axillary (lateral) shoot or root. **2.** To put forth **branches**.

**Braun, A.C. and P.R. White (1943):** Isolated bacteriologically sterile cells in culture, and were the first to demonstrate unequivocally that crown gall resulted from a true neoplastic transformation.

**break:** **1.** A side shoot. **2.** The act or result of **breaking**.

**breeding:** Production of scientifically improved (genetically altered) plant (or animal) varieties.

**5-bromodeoxyuridine (BUdR or BrDU, $C_9H_{11}BrN_2O_5$ mw 307.11):** Growing cells readily incorporate this base analog into their nucleic acids and can then be killed by light exposure. It suppresses the utilization of deoxythymidine. This procedure is a selection mechanism utilized in mutagenic studies to eliminate wild type cells and recover the auxotrophs.

**5-bromouracil (BU, $C_3N_2O_2H_3Br$ mw 190.99):** A base analog closely resembling thymine, used to produce transitional mutations; in which one purine-pyrimidine pair is replaced by another in the DNA.

**browning:** Discoloration due to phenolic oxidation of freshly cut surfaces of explant tissue, a common occurrence. In later stages of culture this discoloration may indicate a nutritional or pathogenic problem, generally leading to necrosis.

**bridge:** A filter paper or other substrate used as a wick and support structure for a plant tissue culture when liquid media are used.

**Büchner funnel:** A porcelain funnel with a perforated flat circular base employed in vacuum filtration. After E. Buchner (1860–1917).

**bud:** **1.** An embryonic or undeveloped, unemerged stem, leaf or flower, often enclosed by reduced or specialized leaves called bud scales. The beginning of incipient growth or development. **2.** To graft onto another type of tree, plant. **3.** A vegetative outgrowth from a yeast.

**budding:** **1.** A general term for the first sign of spring growth. **2.** The process of grafting a stem **bud** to the rootstock of the same or a different variety. **3.** The process by which yeasts produce vegetative outgrowths.

**bud-sport or bud mutation:** A somatic mutation in a bud giving rise to a branch, fruit or flower that is atypical of the plant on which they occur. These characters (if beneficial) may be retained through vegetative propagation from the affected area.

**buffer:** A substance in solution, or system that resists pH change, (despite addition of some small amount of acid or base) and withstands shock; as **buffered** media can resist pH drift. The degree to which pH change is resisted is a measure of the solution's **buffering capacity**.

**bulb:** **1.** An underground food storage or reproductive organ with a short stem on which enlarged and fleshy specialized leaves are developed; as the onion or tulip. **2.** An incandescent lamp.

**bung:** A stopper or closure.

**Bunsen burner:** A hot flame burner using a mixture of gas and air ignited at the top of a metal tube. This device is commonly used for sterilizing tools and container openings during aseptic transfer. After R.W. Bunsen (1811–1899).

**burgeon:** To flourish, bud, sprout or grow.

**burette:** A graduated glass or plastic tube with a stopcock (tap) used to measure or dispense liquid volumes. Hot sterilized media may be dispensed into tubes using this device.

# C

**C:** The chemical symbol for the element **carbon**.

**Ca:** The chemical symbol for the element **calcium**.

**cabinet (growth cabinet):** A cupboard suitable for incubating a small number of culture vessels or tubes under controlled environments. The degree of control over light, temperature or relative humidity is a function of quality of the cabinet but is generally less than that implied by the use of the term incubator or laboratory incubator.

**calcium (Ca aw 40.08 an 20):** A soft, white, abundant metallic element, the chief source of which is limestone **($CaCO_3$)**. A macroelement involved in adhesion of cells to one another to make up the plant body, enzyme activation, movement of plant materials and in root development. It is added in salt form to most nutrient solutions for plant tissue culture, as **calcium chloride, calcium nitrate, calcium phosphate** or **calcium sulphate**.

**calcium alginate:** A hydrophyllic colloid obtained from seaweed. A derivative of alginic acid $(C_6H_8O_6)n$. It reversibly immobilizes plant protoplasts in which condition they can tolerate a reduced osmotic potential.

**calcium hypochlorite ($Ca(OCl)_2.4H_2O$ mw 215):** An algicide, bactericide and fungicide capable of disinfecting and bleaching. It is used in dilute solution (5–10% w/v) as a plant tissue disinfectant often when agitating or using a vacuum. Tissue damage may occur if contact is prolonged. Thorough washing with sterile water generally follows treatment.

**calcium pantothenate or pantothenic acid ($Ca(C_9H_{16}NO_5)_2$ mw 476.54):** The **calcium** salt of vitamin $B_5$. An occasional vitamin additive to plant tissue culture media. Solutions of this vitamin are not stable to autoclaving.

**Calcofluor White ($C_{44}H_{34}N_{12}O_6S_2Na_2$ mw 936.95):** The brand name for a fluorescent dye used in many industrial processes. It binds to cell walls and fluoresces under ultra violet (u.v.) light. It is used in determining the cell wall digestion interval in protoplast formation.

**calliclone:** A clone derived from **callus**. A high degree of genetic and phenotypic variability is associated with **calliclones**.

**callus, pl. calli or calluses: 1.** Wound tissue, tissue formed on or below a wounded surface. **2.** Disorganized tumor-like masses of plant cells that form in culture. These proliferate in an irregular tissue mass, and vary widely in texture, appearance and rate of growth; a function of the tissue type (species and explant) and the composition of the medium. The process of callus formation is **callogenesis**.

**callus culture:** The cultivation of callus, usually on solidified medium and initiated by inoculation of small explants or sections from established organ or other cultures (inocula). Callus may be maintained indefinitely by regular subdivision and subculture. It may be used as the basis for organogenetic (shoot, root) cultures, cell cultures or proliferation of embryoids.

**calomel:** See **mercuric chloride**.

**cambium:** A lateral meristem (persistently meristematic tissue) in the stems and roots of dicots giving rise to parallel rows of cells (secondary tissues); increasing the girth of plant organs. This term applies to both the vascular cambium and the cork cambium or phellogen. In the **cambial** zone between the secondary phloem and secondary xylem are thin-walled meristematic cells, the fusiform and ray initials and their recent derivatives. The fusiform initials divide periclinally forming radially oriented files of secondary xylem and phloem cells. The ray initials are isodiametric and divide to form parenchymatous vascular rays. The cork cambium divides periclinally forming radially seriate cork or phellem to the outside and secondary cortex or phelloderm to the inside.

**cannula:** A small tube for insertion into tissue; as a cork borer.

**cap:** Cover or closure.

**Caplin, S.M. and F.C. Steward:** These men demonstrated (1940's, 1950's) the promotive effect of coconut milk used with synthetic auxins for previously difficult to grow tissues and species.

**capsid:** The protein coat of a virus particle, surrounding and protecting its nucleic acid. Serological reactions of capsid proteins are used in virus identification.

**carbohydrate:** One of a large group of organic compounds composed of carbon, hydrogen and oxygen including the monosaccharides and disaccharides (sugars) and the polysaccharides (starches, cellulose and others). Many have the structure $C(H_2O)_n$. These compounds are essential in metabolism. Cellulose is the main plant structural component and starch the main form of plant stored food. A carbohydrate source is commonly included in plant tissue culture media (sucrose, glucose, etc.), providing an energy source and controlling the osmotic environment.

**carbolfuchsin (Ziehl's stain):** A mixture of **fuchsin** ($C_{20}H_{19}N_3HCl$ mw 337.8), a red dye, in ethanol and aqueous phenol. It is used in many microscopical stains. It stains chromosomal material and is used in chromosome analysis as a differential nuclear stain to verify somatic cell hybrids.

**carbon (C aw 12.011 an 6):** An amorphous or crystalline element. Its atoms have a valence of four; can unite with each other forming the very large molecules on which life on earth is based. It occurs in many allotrophic forms and many organic compounds. An essential major element obtained by plants from nutrient medium carbohydrate or from photosynthetically fixed **carbon dioxide** in the air.

**carbon dioxide ($CO_2$ mw 44.01):** An incombustable compound gas heavier than air, odorless and colorless. It makes up 0.033% (by volume) of the atmosphere. Plants use it in photosynthesis and plants and animals return it to the air by respiration. At −78.5°C it exists as dry ice and is used as a refrigerant.

**carbon source:** A source of the non-metallic element carbon (C); as organic substances like sugars taken up and metabolized by plant tissue cultures.

**carborundum:** A hard black crystalline solid, of silicon carbide (SiC) from heating silica ($SiO_2$) with carbon in a furnace. It is used for its abrasive, refractory or adsorbtive qualities.

**carbowax:** See **polyethylene glycol**.

**carboxydismutase:** See **ribulose bisphosphate carboxylase**.

**carcinogen:** A substance capable of inducing a cancer (teratoma, carcinoma).

**Carnoy's fixative:** Consists of 6 parts ethanol:3 parts chloroform:1 part acetic acid. It is used in chromosome analysis.

**carotene or carotin ($C_{40}H_{56}$ mw 536.85):** A hydrocarbon group of yellow, orange and red pigments synthesized by plants. They act as photosynthetic accessory pigments; gathering light and screening (protecting) other photosynthetic pigments.

**carotenoid:** One of a group of fat soluble pigments associated with chlorophylls which includes **carotenes** and their oxygenated derivatives, the xanthophylls. **Carotenoids** are yellow, orange and red pigments located in the chloroplasts and chromoplasts of plants acting as photosynthetic accessory pigments.

**carpel:** The angiosperm gynoecium, composed of the stigma, style and ovary.

**carrot:** See *Daucus carota*

**casamino acid:** An amino acid obtained through the digestion (acid hydrolysis) of the milk protein **casein**. Protein digest yields **casamino acids** and other substances. These are undefined constituents of some plant tissue culture media.

**casein:** The principal protein of milk, a phosphoprotein.

**casein hydrolysate (edamin):** A milk protein digest product composed of amino acids (casamino acids) and other substances. This complex (undefined) product is sometimes used as an addition (0.02-0.10%) in nutrient solutions used in plant tissue culture as a non-specific source of organic nitrogen.

**catalyst:** A compound or agent which enhances the rate of a chemical reaction but is not itself changed in the process. Enzymes are organic **catalysts** produced by cells. Catalysts function through transient combination with reactants to form a transitional complex with lower activation energy than that of uncatalyzed reactions. The process is **catalysis**.

**cation:** A positively charged **ion** in solution.

**caulogenesis:** Shoot formation; as de novo shoot development from callus.

**caustic soda:** See **sodium hydroxide**.

**cell:** The fundamental structural and physiological unit of the bodies of plants and animals. An organized unit of protoplasm usually enclosing a central nucleus and surrounded by a membrane. In addition plant cells possess a rigid outer cell wall. In many plant tissues this cell wall persists after the cell ceases to live.

**cell count:** The number of cells per unit suspension volume or callus weight. Tissue is treated with chromic acid (5–8%) or pectinase (0.25%) for up to 15 minutes followed by mechanical dispersion, then cell numbers are estimated with a hemocytometer (haemocytometer).

**cell culture:** The culture of single or groups of cells on solid or dispersed in liquid nutrient media (cell suspension cultures) usually following a set of defined growing conditions (protocols).

**cell division:** Nuclear division (karyokinesis) followed by division of cytoplasm (cytokinesis) into two equal parts (daughter cells), each with a nucleus, by the formation of a cell plate. In mitosis, each daughter cell has the same number of chromosomes as the mother cell. In meiosis, after reduction division the daughter cells have half the chromosome number of the mother cell.

**cell hybridization:** Formation of synkaryons, viable cell **hybrids** produced through cell fusion. They are identifiable by their increased chromosome number compared to the parent cells and the possession of characters found in one or another of the parental cells.

**cell line:** Developmental history or descent, through cell division from a single original cell; as callus may constitute a group of cells, all descendants from a single cell plating. Numerous **cell lines** may be present in a culture. Any deviation in culture technique may favor one cell line over another.

**cell membrane:** Usually synonymous with the plasma membrane or plasmalemma; the surface layer of a plant cell beneath the cell wall, or may include the membrane surrounding any cellular organelle. They are characterized by selective permeability.

**cell number:** The absolute number or approximation of the number of cells per unit area of a culture or medium volume.

**cell plate:** The structure developing during telophase (mitosis) at the equatorial plane of a plant cell between the daughter nuclei. The plate consists of small membrane-bound vesicles which fuse, forming the middle lamella and then the primary cell walls, as the cell divides into two daughter cells. Channels left between fusing vesicles become plasmodesmata. The plate unites with the walls and plasmalemma of the mother cells, completing cell division.

**cell selection:** Selection within a group of genetically different cells; involves competition between cells often under some stress. Selection criteria may involve cell viability; a unique phenotype or biochemical activity; or another basis for choice. Select cells or cell lines are generally relocated to fresh medium for continued selection and often are exposed to an increased level of the stress agent. The final objective is usually to regenerate plants from those select cells, in hope that the plants will exhibit the traits selected for at the cellular level.

**cell strain:** A cell line characterized by unique and consistent feature(s).

**cell suspension:** Cells and small aggregates of cells suspended in a liquid medium; as in cell suspension cultures. Explants, or callus derived from them are transferred to liquid medium and the cultures are then agitated on a mechanical shaker. The ensuing single cells and small cell clusters are used for a number of purposes in plant tissue culture; as in single cell cloning.

**cell wall:** The limiting layer of a plant cell, structurally rigid and mechanically supportive. It is nonliving, cellulosic tissue traversed by plasmodesmata which are delicate protoplasmic connections. Primary cell walls are composed of cellulose, hemicellulose and other polysaccharides. They are relatively elastic and extensible, and must grow as the cell grows. Secondary cell walls are composed of fewer hemicelluloses, and include lignin or other substances. These layers provide mechanical strength and are present in cells that may not be living at maturity. In protoplast formation cell walls are removed using mechanical methods or enzyme preparations.

**cellobiose ($C_{12}H_{22}O_{11}$ mw 342.17):** A disaccharide of glucose units linked by a glycosidic bond. An intermediate product in the hydrolysis of cellulose by the enzyme cellulase.

**cellulase or 4-glucanohydrolase:** An enzyme complex that hydrolyzes cellulose to the smaller fragments, cellobiose, by degrading $\beta$(1–4) linkages. These in turn are hydrolyzed by cellobiase to glucose. Cellulase is used alone or with other enzymes to digest plant cell walls to produce protoplasts.

**cellulose ($(C_6H_{10}O_5)_n$ mw $(162.14)_n$):** An insoluble complex carbohydrate (polysaccharide) made of microfibrils (chains) of hundreds of $\beta$(1–4) linked glucose molecules. Cellulose is the structural component of plant cell walls. It is synthesized in the golgi apparatus.

**Cellulysin:** A brand name for cellulase isolated from *Trichoderma viride*; an enzyme used to degrade cellulose in the cell wall digestion process used in protoplast formation.

**central mother cell:** A large, vacuolated, subsurface cell in a shoot apical meristem.

**centrifuge:** An apparatus used for separating particles from suspension using centrifugal force. Balanced tubes containing the suspension are rapidly rotated causing sedimentation of particles to the bottom of the tubes (pelleting) based on their weight and the density of the suspending medium.

**centrosome or microcentrum:** Zone of differentiated centriole-containing cytoplasm. Not present in higher plants.

**certify:** To attest, vouch for, assure or endorse as meeting a set required standard; as **certified** (pathogen tested) plant material or certified seed is guaranteed free of the pathogens it has been specifically tested for.

**centriole:** In mitosis this small spherical body forms the center of the astral rays.

**chalaza:** The base of the ovule, where the integuments and nucellus are fused, and often to which the funiculus is attached; as the **chalazal** end of an ovule explant may be severed from its attachment site on the ovary and the ovule placed into culture (ovule culture).

**character:** The phenotype of an individual organism; the resultant of specific environmental influences on the genetically determined reaction norm of its genotype. Also refers to a single trait or structure.

**charcoal:** One of many varieties of carbon. It is black, porous, imperfectly combusted organic matter, like burned wood. It has adsorbent and filtering qualities. See **activated charcoal**.

**chelate:** A chemical compound with which metal atoms may be combined in order to prevent them from precipitating out of solution and so being unavailable to plants; as ethylenediamine tetra-acetic acid, disodium salt ($Na_2EDTA$) is complexed to iron (and other metal ions) in nutrient solutions used for plant tissue culture. The process is **chelation**.

**chemically defined medium:** A nutrient medium for plant tissue culture in which all of the chemical components are fully known and stated. Quality (high purity) chemicals are utilized, and no undefined constituents are included.

**chemical mutagen:** A chemical capable of causing genetic mutation in the organisms exposed to it.

**chemostat:** An open, continuous culture system whereby the inflow of fresh medium including a growth limiting compound is monitored to maintain constant culture growth. Balancing the fresh medium inflow is a regulated outflow of cells and spent medium.

**chemotherapeutant:** A chemical used to pretreat diseased source plants prior to excision or incorporated into media to support some therapeutic objective; as malachite green or virazole (ribavirin) are used to eliminate virus in meristem tip culture. The process is **chemotherapy**.

**chimera or chimaera:** A plant or tissue composed of more than one kind of genetic tissue. In a periclinal chimera one genotype occurs in a superficial layer, covering a genetically different core. In a mericlinal chimera one genotype occurs in a localized part of the plant and the rest is occupied by another genotype. In a sectorial chimera two genotypes share distinct sectors of the plant. Chimeras can occur naturally; be established artificially by grafting; or developed through somatic mutation by chemicals such as colchicine, or by other means in plant tissue cultures.

**chlorenchyma:** The general term for chloroplast-containing parenchyma cells; as leaf mesophyll tissues.

**chlorine (Cl aw 35.453 an 17):** An abundant elemental halogen gas with a greenish yellow color and a suffocating odor. It is widely used in bleaches and disinfectants for its germicidal properties. It forms complexes with a wide range of elements. It is a microelement component of mineral salts used in nutrient solutions for plant tissue culture.

**(2-chloroethyl)phosphonic acid:** See **ethephon**.

**4-chlorophenoxyacetic acid or para-chlorophenoxyacetic acid (pCPA, $C_8H_7O_3Cl$ mw 186.59):** A synthetic hormone of the auxin type which is sometimes used in plant tissue culture media.

**chlorophyll:** The green plant pigment necessary for light capture in photosynthesis. Chlorophyll molecules consist of a magnesium-porphyrin "head", and a long phytol "tail". Chlorophyll is found in higher plants, usually in the thylakoids of chloroplasts, in some protista and in some prokaryotes. There are ten or more **chlorophylls**, differing in minor ways chemically. Only **chlorophyll a ($C_{55}H_{72}O_5N_4Mg$)** is common to all oxygen-evolving photosynthetic organisms.

**chloroplast:** A cytoplasmic, double membrane bound organelle (plastid) containing chlorophyll and other pigments in photosynthetic eukaryotic cells. Inside, embedded in the stroma (matrix) are sac-like vesicles (thylakoids) forming grana stacks in higher plants, the sites in which light energy is converted to chemical energy in photosynthesis.

**chlorosis:** Reduced development of or loss of chlorophyll, leading to yellowing or whitening of normally green tissue. It is caused by reduced light (etiolation); water; mineral deficiency; disease; genetic factors (albinism or chimeral growth); or for other reasons. Such plants are **chlorotic**; their chloroplast size or numbers are reduced or chlorophyll has been destroyed or its synthesis affected.

**choline chloride ($C_5H_{14}ClNO$ mw 139.63):** An alkaloid, (B complex vitamin), occasionally added to plant tissue culture media. **Choline**, found in many plant organs, is the basic constituent of lecithin.

**chromatid:** One of two chromosome filaments (daughter chromosomes) joined by a centromere during mitotic interphase, and separating during anaphase. Mitotic **chromatids** are identical but meiotic chromatids, due to crossing over, may not be identical.

**chromic acid or chromium trioxide ($CrO_3$ mw 100.01):** This compound exists only in solution and is made by mixing potassium dichromate and concentrated sulphuric acid. A dilute solution may be used to prepare tissues for hemocytometer (haemocytometer) counts to determine cell numbers in cell suspension cultures. Concentrated solutions were once used for cleaning new or dirty glassware but have now been supplanted by detergents.

**chromoplast:** A pigment-containing colored plastid; as is a chloroplast, or one in which carotenoids predominate.

**chromosome:** A structural nuclear unit of double stranded DNA containing many genes. These store and transfer genetic information within the nucleus of eukaryotic cells. The number, size and shape of **chromosomes** at prophase are typical of the species. They replicate during each cellular division (pair of chromatids). In micropropagation it is not desirable to alter these in any way. A common objective of mutagenic studies is to change them in some beneficial way.

**cion:** Obsolete term, see **scion**.

**citric acid or citrate ($C_6H_8O_7$ mw 192.12):** A white crystalline acid of wide distribution in plant and animal tissues. It is present as a free acid in citrus and in other sour fruits. It has uses which include those of antioxidant and sequestering (chelating) agent for trace metals. It is commonly included in disinfecting solutions and in some tissue culture media.

**citric acid cycle:** See **tricarboxylic acid cycle**.

**Cl:** The chemical symbol for the microelement **chlorine**.

**cladode or cladophyll:** A stem which resembles a leaf in appearance.

**cline:** The gradation in phenotypic (**phenocline**) or genotypic (**genocline**) characteristic(s) within a population of a species over a geographical or ecological distribution range.

**clonal multiplication or clonal propagation:** Vegetative (asexual) propagation from a single cell or plant. See **clone**.

**clone:** **1.** A basic category of cultivar. The original cultivar or variety. Not to be confused with source clone, indicating different origins within the clone. Members have a common origin, are the extension of a single cell (via mitosis) or plant and have been produced by vegetative means only. Genetic uniformity is accepted. A clone may have one or more sources. **2.** While the expectation is that members of a clone are both phenotypically and genotypically identical, this assumption is not always true and members may not be genetically or phenotypically equivalent (variants). The objective of much tissue culture propagation is to propagate very large numbers of selected plants with the same genotype. The process is **cloning**. The prefix meri reflects the explant used to initiate the clone (meristem tip source clone). In a **calliclone** cloning involves the callus stage. **3.** Specific DNA sequences are said to be **cloned** when isolated and propagated.

**closed continuous culture:** A cell suspension culture with a continuous influx of fresh medium, maintained at constant volume by the efflux of spent medium. All cells are retained within the unit. See **continuous culture**.

**closure:** A cap or cover to shield, close or enclose a test tube, jar or other container. It may be metal, plastic, cotton or of other material.

**Co:** The chemical symbol for the element **cobalt**.

**cobalt (Co aw 58.9332 an 27):** A pinkish grey microelement, the salts (**cobaltous chloride** or **cobaltous nitrate**) of which are commonly used in plant tissue culture media.

**Cocking, E.C. (1960):** Obtained the first higher plant protoplasts using root cells subjected to fungal cellulase and demonstrated new cell wall regeneration on protoplasts from tomato fruit locule tissue.

**coconut milk or coconut water:** Liquid endosperm from the center of the coconut seed. A complex, undefined addendum of variable quality and effects in some nutrient solutions (2–15%, v/v) for plant tissue culture. It has growth promoting effects and cell division factors. It is replaceable in some cases by cytokinins and/or sugar. It was first used by van Overbeek et al. (1941) to stimulate *Datura* embryo cultures.

**co-culture:** The joint culture of two or more types of cells; as a plant cell and a microorganism or two types of plant cells; as is done in various dual culture systems or the nurse culture technique.

**codon:** The mRNA basic coding unit, a triplet of nucleotides, determining the incorporation of a particular amino acid into a polypeptide chain. On the tRNA is a complementary triplet of nucleotides, the **anticodon**. The specificity for translating genetic information from DNA into mRNA, then to protein, is provided by codon-anticodon pairing.

**coenocyte:** An organism or a portion thereof that is multinucleate; the nuclei are not each separate in one cell; as some protoplast or cell fusion products are multinucleate or **coenocytic**.

**coenzyme:** An organic molecule associated with an enzyme and essential to some enzyme-catalyzed reactions. **Coenzymes** usually function as temporary carrier molecules in the removal or transfer of atoms, electrons or functional groups. In some cases the coenzyme is loosly bound to the enzyme. A coenzme is termed a prosthetic group when it is tightly bound to the enzyme. Important coenzymes include NAD and NADP.

**colchicine ($C_{22}H_{25}NO_6$ mw 399.43):** A poisonous (alkaloid) drug obtained from the meadow or autumn crocus (*Colchicum autumnale*), used to arrest mitosis at metaphase and induce chromosome duplication without separation, resulting in diploidization of haploid cells (homozygous diploidization) and polyploidization of diploid cells.

**coleoptile:** The sheath enclosing and protecting the epicotyl in a grass embryo. It is sometimes called the first leaf.

**coleorhiza:** The sheath enclosing and protecting the radicle in a grass embryo.

**collenchyma:** Support cell or tissue in regions of primary growth in some plants, usually beneath the epidermis. Collenchyma cells are elongated, alive at maturity and have irregularly thickened primary cell walls capable of extension.

**colleter:** A multicellular glandular hair or trichome.

**colony:** **1.** A group of interdependent cells or organisms. **2.** An aggregate of cells developed from a single cell; as in single cell platings, a clone composed of one cell line.

**companion cell:** See **phloem**.

**compensation point:** **1.** The light intensity at which the amount of carbon dioxide ($CO_2$) taken up by plants during photosynthesis is equivalent to that released during respiration. **2.** That concentration of $CO_2$ at which $CO_2$ release from respiration and uptake during photosynthesis are equal.

**competent:** Able to function or to develop; as embryogenically competent cells are capable of developing into fully functional embryos. The opposite is **non-competent** or morphogenetically incapable.

**complex substance:** A complicated and undefined substance; as in some addenda to nutrient solutions used in plant tissue culture. Examples include protein hydrolysate, yeast or malt extracts, endosperm from corn or coconuts, juices like orange or tomato juice and others.

**conditioning:** **1.** A phenotypic alteration resulting from the action of external agents during critical developmental stages. **2.** An undefined interaction between media and tissues encouraging the growth of single cells or small aggregates. **Conditioned** medium may be used to induce growth of cells plated at low densities (below the minimum inoculum size) or cells unable to grow for unknown reasons. Conditioning may be accomplished by immersing cells or callus contained within a porous material (such as dialysis tubing) into fresh medium for a time dependent on cell density and a volume related to the amount of fresh medium.

**conductivity, electrical:** The electrical conducting capacity of water due to dissolved salt (ion) content. Conductivity is measured quantitatively per unit area, per unit potential gradient per unit time. It increases as water purity decreases and is the reciprocal of resistivity. It is expressed as micro- or milli-mho/cm (specific conductance).

**conifer:** Any of a large order (**Coniferales**) of temperate region **cone**-bearing, mostly evergreen forest trees and shrubs, like pine, spruce, larch, cypress and others.

**contaminant:** An undesirable bacterial, fungal or algal microorganism accidentally introduced to a culture or culture medium. It may overgrow the plant cells or inhibit their growth through release of toxic metabolites. Rigorous exclusion of potential **contaminants** must be practiced in plant tissue culture.

**contaminate:** To accidentally introduce a substance or organism (**contaminant**) into a medium or culture. The process is **contamination**.

**continuous culture:** A cell suspension culture with a continuous influx of fresh medium, maintained at constant volume by the efflux of spent medium (closed continuous) or with the efflux of cells and spent medium (open continuous).

**control:** **1.** The untreated plant or unchanged (standard) protocol or treatment for comparison with the experimental treatment. **2.** To direct or regulate; as in induction of organogenesis in cultures through hormone regulation.

**controlled environment:** A chamber, room or situation in which the environmental parameters of light, temperature, and relative humidity are controlled. The partial pressure of gases may also be controlled.

**convergence:** **1.** Evolution of or environmentally induced similarity in some characteristic(s) between initially different groups of cells or organisms. **2.** The process of approaching or coming to a limiting value or a point.

**co-ordinate system:** Used to locate and relocate small items in relatively large areas; as original single cell inoculum sites. A T-shaped marker on the bottom of the petri dish facilitates precise placement of the dish over a labeled grid.

**copper (Cu aw 63.54 an 29):** A reddish metallic microelement. **1.** A constituent of some respiratory enzymes and essential in chlorophyll synthesis and photosynthetic electron transport. Copper salts (**copper chloride**, **cupric nitrate** or **cupric sulphate**) are added to nutrient solutions for plant tissue culture. **2.** A maleable and ductile, electrical conductor.

**cork or phellem:** Secondary tissue formed by the cork cambium (phellogen). It is composed of polygonal, suberized cells which are non-living at maturity. It forms the protective external tissue of stems and roots which replaces the epidermis and is impermeable to water and gases. Stoppers and similar items are made from the bark of the cork oak (*Quercus suber*).

**cork borer or cannula:** A sharp-edged metal cylinder used to punch out discs or cores of plant tissue. The latter are subsequently sliced into discs or wedges. This is a common means of acquiring explants for plant tissue culture.

**cork cambium or phellogen:** A lateral meristem forming the periderm producing cork (phellem) to the outside and secondary cortex (phelloderm) to the inside of stems and roots of dicots and gymnosperms.

**corm:** An underground food storage or reproductive organ comprised of a thickened stem, leafless or with poorly developed scale-leaves. Buds form in the axils of these scales, the remnants of the previous years' growth. Multiplication is through **cormels** which are miniature **corms**. Examples are gladiolus and crocus.

**corn milk:** Corn endosperm tissue. An undefined complex addendum to some plant tissue culture media.

**corolla:** A collective term for the petals of a flower.

**corpus (LIII):** The cell group below the two tunica layers (LI, LII) in the shoot apex. These divide in both anticlinal and periclinal planes, growing in volume. The tunica corpus theory suggests that the corpus gives rise to the internal plant body, while the tunica layers form the external plant including the epidermis.

**cortex:** The ground tissue (primary tissue) region, primarily of parenchyma cells located between the epidermis and the stele of stem or roots.

**cotransformation:** The transfer of a well characterized plant trait, expressed in vitro, which is linked to another plant trait or traits not expressed in culture.

**cotton or cotton wool:** A soft, fibrous, white material obtained from the cotton plant (*Gossypium hirsutum*). Nonabsorbent cotton is commonly used to stopper flasks and is usually wrapped in several layers of cheese cloth or muslin. Burning or singing cotton wool emits vapors toxic to living cells.

**cotyledon:** The (seed) leaf or leaves of the embryonic plant; one in the **monocotyledons**, two in the **dicotyledons** and a variable number in the gymnosperms. These leaves generally have a food storage function in dicotyledons, and absorb food from the endosperm for the embryo in monocotyledons.

**cotyledonary node:** The point of attachment of the cotyledons to the embryonic axis.

**critical concentration:** The chemical concentration above or below which a reaction or developmental process will not proceed.

**cross:** The process or product of interbreeding at the cell or organism level.

**cross over:** Nuclear exchange between homologous chromatid portions in meiosis. Resultant recombination generates new genetic variation through breakdown of existing linkage groups.

**crown:** **1.** The stem-root junction (base) of a plant, especially an overwintering base of a herbaceous perennial. **2.** The tree-top.

**crown gall:** A disease of plants in which tumors form. The causal agent is the bacterium, *Agrobacterium tumefaciens*. Gall tissue has been grown in vitro. A tumor inducing portion of the bacterial genome (the Ti plasmid) may be used experimentally as a genetic tool to incorporate (vector) genetic information into plant cells.

**crucifer:** A plant belonging to the Cruciferae (Brassicaceae) or mustard family, a large dicotyledonous family of important crop and ornamental plants (turnip, cabbage, cauliflower, etc.).

**cryopreservation or freeze preservation:** To freeze cells in the presence of protective (**cryoprotective**) agents and store them in the frozen state until required.

**cryoprotectant:** An agent able to prevent freezing and thawing damage to cells as they are frozen or defrosted. These substances have high water solubility and low toxicity. They are classified either as permeating (glycerol and dimethyl sulfoxide) or non-permeating (sugars, dextran, ethylene glycol, polyvinyl pyrolidone and hydroxyethyl starch) agents.

**Cu:** The chemical symbol for the microelement **copper**.

**cultivar or cultivated variety:** A group of cultivated plants (sometimes a clone) distinguished from the species by some horticultural or agricultural character(s) that is (are) perpetuated when the plants are propagated either sexually or asexually.

**culture:** **1.** A general term for the **cultivation** of microorganisms, animals, plants or their cells in vivo or in vitro especially in order to improve the breed. **2.** In vitro aseptic cultivation of microorganisms or of cells, tissues, or organs of plants, animals in or on prepared nutrient media under controlled, aseptic conditions.

**culture alteration:** A persistent change in a cultured cell or tissue anatomy or physiology, including a change in one or more nutritional requirements or a change in its proliferative capacity.

**culture medium:** A prepared nutrient solution (substrate) that may be chemically defined for growing plant tissues or other organisms in vitro.

**culture room:** A controlled-environment room (light, temperature, relative humidity, often air conditioned, etc.) for incubation of plant cultures.

**culture tube:** A test tube used to contain medium and an explant or culture derived from it, usually incubated in a controlled environment.

**culture vessel:** A container used to hold medium and an explant or culture derived from it, usually incubated in a controlled environment.

**cuticle:** A waxy or fatty membraneous protective layer (noncellular) of **cutin**, a fatty substance, secreted by and covering the epidermis and protecting plants from excessive water loss. It may be affected (qualitatively or quantitatively) by the high relative humidity of culture. Its formation is **cuticularization**.

**cutting:** An explant (small or large) **cut** for propagative purposes.

**cyanocobalamin ($C_{63}H_{88}CoN_{14}O_{14}P$ mw 1355.64):** Known as vitamin $B_{12}$, it is an occasional additive in plant tissue culture media which is nonessential but perhaps stimulatory to plant growth.

**cybrid:** A cell or plant cytoplasmic hybrid (heteroplast) with the nucleus of one and cytoplasmic organelles of another, or of both cells or plants.

**cysteine ($C_3H_7NO_2S$ mw 121.13):** An amino acid in proteins and precursor of coenzyme A and cystine.

**cysteine hydrochloride ($C_3H_7NO_2S.HCl$ mw 157.63):** An occasional, amino acid (cysteine) salt additive in plant tissue culture media. It is included primarily for its antioxidant properties.

**cytochimera or chromosomal chimera:** Cells within a tissue possessing different chromosomal numbers, as when a callus consists of diploid and polyploid regions of cells.

**cytodifferentiation:** A functional and/or morphological differentiation occurring in cells during ontogenesis, affecting their phenotype.

**cytogenetics:** The study of the hereditary component of cytological changes, primarily chromosomal effects.

**cytokinesis:** The division of the cytoplasm of a cell into two or more daughter cells in meiosis or mitosis. In conjunction with nuclear division (karyokinesis) the term means cell division.

**cytokinin:** One of a large class of plant hormones (phytohormones) or synthetic growth substances. Cytokinins and many of their analogs have been chemically synthesized. They promote cell division and enlargement in cultures in the presence of auxins and have other effects such as control of organ (bud, root) differentiation; influence auxin transport; inhibit senescence, abscission of leaves and breaking of dormancy and apical dominance in buds. They resemble kinetin (K) (the type member) in physiological activity. They are $N^6$-substituted aminopurine compounds. In tissue culture, these hormones are employed to stimulate cell division and induce axillary bud proliferation. Cytokinin withdrawal stimulates rooting of plantlets. The most common **cytokinins** used in plant tissue culture are K (*N*-(2-furanylmethyl)-1*H*-purin-6-amine), *N*-(phenylmethyl)-1*H*-purin-6-amine or 6-benzylaminopurine (BA or BAP) and N-(3-methyl-2-butenyl)-1*H*-purin-6-amine or 2-isopentyladenine (2-iP). Stock solutions of these are prepared by dissolving in HCl acid (ca. 1 M) then making up to volume with water. These stock solutions are usually refrigerated.

**cytolysis:** Cell dissolution or disintegration.

**cytoplasm:** The protoplasm, all the living material of a cell, exclusive of the nucleus. **Cytoplasmic** pertains to the cytoplasm.

**cytoplasmic hybrid or cybrid:** A cell or plant hybrid (heteroplast) with the nucleus of one and the cytoplasmic organelles of another or of both cells or plants.

**cytoplasmic inheritance:** The transmission of hereditary characters via the DNA of extranuclear organelles or plasmids. The expression of these cytoplasmically determined characters is not related to chromosome behaviour or movement and consequently does not segregate in Mendelian ratios. These characters are transmitted through the female gamete, which contributes most of the cytoplasm to the zygote, including small quantities of nuclear material found in chloroplasts, mitochondria and sometimes the cytosol. An example of cytoplasmic inheritance is a form of male sterility in maize.

**cytoplasmic variant:** A maternally inherited change in a cellular trait.

**cytotoxic:** A chemical or other agent toxic to cells.

# D

**2,4-D:** See **(2,4-dichlorophenoxy)acetic acid**.

**de novo:** (Latin: "from the beginning, anew"). Arising, sometimes spontaneously, from unknown or very simple precursors.

**deceleration phase:** The declining growth rate phase following the linear phase and preceeding the stationary phase in most batch suspension cultures.

**deciduous:** Falling off or shed at maturity or seasonally; as leaves, petals or fruit. Characterized by such a falling off; not evergreen.

**decontaminate:** To free from **contamination** or surface sterilize.

**dedifferentiation:** The resumption of meristematic activity by more or less mature cells through a reversal of the process of cell or tissue **differentiation**. Cell division leading to the formation of small, microvacuolate, isodiametric cells with prominent nuclei, that are capable of organized development; as when adventitious buds or roots develop on mature tissue.

**deficiency: 1.** An insufficient supply or unusable form, of one or more nutritional or environmental requirements, for plant development, growth or physiological function. The absence of adequate conditions for plant growth or performance, resulting in disease. **2.** The deletion of a gene or a series of genes. This may result in a mutant phenotype or may be lethal.

**defined: 1.** Precisely stated and fixed conditions of medium, environment and protocol for plant growth. **2.** Fully known and stated components of plant tissue culture medium.

**dehumidifier:** An apparatus that removes moisture from the air.

**deionized water:** Water that has been passed through an ion exchange device to remove soluble minerals and some organic salts. The process is **deionization**.

**deletion: 1.** Omission, removal or cancellation of some factor. **2.** The lost portion of a chromosome or a nucleotide sequence in nucleic acid.

**demineralize:** To remove the mineral content (salts, ions); as from water. Removal methods include ion exchange, distillation and electrodialysis.

**deoxyribonucleotide:** A nucleotide that contains, as a pentose component, 2-deoxy-D-ribose.

**deoxyribonucleic acid (DNA):** The DNA molecule is unique and intricate; up to several million nucleotide units form a double helix containing 2-deoxyribose, phosphoric acid and the nitrogenous bases adenine, guanine, cytosine and thymine. This complex polymer of sugars and nucleoproteins in a specific sequence carries the hereditary information in chromosomes.

**dermal:** Of or pertaining to **epidermal** or **peridermal** tissue.

**dermatogen:** In the histogen theory, the plant apical meristem tissue from which the epidermis is derived.

**derepression:** The mechanism by which repression is alleviated; as when a **repressing** metabolite is removed, resulting in the increased level of a protein or enzyme.

**derivative:** Resulting from or characterized by decent through meristematic cell division.

**descendant:** An individual resulting from the sexual reproduction of a pair or successive pairs of individuals.

**desiccate:** To dry, exhaust or deprive of water or moisture. A chemical used for this purpose is a **desiccant**. An apparatus for drying and preventing hygroscopic samples from rehydrating is a **desiccator**. The process is **desiccation**.

**detached meristem:** An axillary meristem or meristem distant from the apical meristem and giving rise to axillary buds and shoots.

**detergent:** A cleaning agent; one of numerous synthetic cleaning preparations chemically different from soap. **Detergents** are commonly used to prepare work areas for aseptic plant tissue culture manipulations and to remove dirt and microorganisms from plant material prior to explantation.

**determinate growth:** Growth of limited, often set duration, of fixed fate; as in most floral meristems and leaves. The direction of differentiation becomes irreversibly established. This process contrasts with the usual culture growth which is (theoretically) infinite, and so indeterminate.

**determination or topophysis:** **1.** The process of commitment to a specific pathway of development (cell, tissue, organism, growth form). This commitment may occur well in advance of the appearance of the phenotype. The perpetuation of the phenotype provides an example of the stability of this process. **2.** A phenomena in which explanted meristems and ensuing growth derived from them in culture, taken from different areas of a plant representing different phases, perpetuate the different phase phenotypes of that plant; as meristems from thorny (juvenile) citrus give rise to thorny individuals unlike meristems explanted from adult tissue that perpetuate the thornless condition in propagules derived from them.

**development:** A process of regulated change (growth and differentiation) of an organism or its parts from its origins towards maturity. This process has evolutionary implications.

**deviation:** An alteration or departure from the typical plant form, function or behavior. It may result from mutation or stress that shifts a developmental process in an alternative direction.

**Dewar flask:** A double-walled glass flask used to keep liquids at other than ambient temperatures; as a thermos does. The inner wall is silvered and a high vacuum is maintained between the two walls to minimize heat transfer to or from the sample chamber. Named after J. Dewar (1842–1913).

**dextrose:** See **glucose**.

**diageotropism:** A gravitationally-induced growth response, where rhizomes are aligned at right angles to the direction of gravitational force.

**dicamba or 3,6-dichloro-2-methoxybenzoic acid ($C_8H_6O_3Cl_2$ mw 221.04):** A synthetic growth regulator of the auxin type with herbicidal properties. It is used to promote in vitro callus growth and as a herbicide. It is dissolved in base (ca. 1 M KOH or NaOH).

**(2,4-dichlorophenoxy)acetic acid (2,4-D, $C_8H_6O_3Cl_2$ mw 221.04):** A synthetic auxin. It is commonly employed as a component of weed killers and in plant tissue culture media to promote callus growth. It is dissolved in base (ca. 1 M KOH or NaOH) but is not very soluble. Alcohol is often used to dissolve 2,4-D. However, many tissues are sensitive to alcohol at low concentration.

**3,6-dichloro-2-methoxybenzoic acid:** See **dicamba**.

**dicotyledon or dicot:** The larger of two angiosperm (flowering plant) classes (**Dicotyledoneae**) with over 150,000 species of trees, shrubs and herbaceous plants. Plants in this class have an embryo with two cotyledons, pinnately or palmately veined leaves, flower parts in fours, fives or multiples thereof, a persistent primary root that develops into a taproot, a cambium and the stem vasculature is a ring of open bundles.

**differentiate:** To undergo physiological and morphological change in cells, tissues or organs as they develop from relatively unspecialized (**undifferentiated**) to the more specialized roles and functions. Refers to the formation of cells, organs or other structures from cells or callus in vitro. The process is **differentiation**. The reverse process is **dedifferentiation**.

**diffusion:** The net movement of molecules from a region of high free energy to a region of lower free energy. In solutions dissolved particles tend to become distributed from regions of high concentration and at equilibrium are evenly distributed.

**digest:** To convert complex, usually insoluble materials to simple often soluble units through enzymatic (**digestant**) action; as in the removal of plant cell walls during protoplast formation. The process is **digestion**.

**dihaploid or heterozygous diploid:** The product of crossing doubled monoploids.

**dihybrid:** A **hybrid** whose parents differed in two pairs of heterozygous genes.

**dimethyl sulfoxide (DMSO, $C_2H_6OS$ mw 78.13):** A highly hygroscopic liquid with little odor or color. It is an organic cosolvent sometimes used in small quantities to dissolve neutral organic substances in plant tissue culture media preparation. Readily penetrates skin. An extremely powerful solvent. It can usually be replaced by a less toxic solvent. It also has uses as a cryoprotectant.

**dimorphism:** The existence of two morphological forms in one plant or in one species; as juvenile and adult phase leaves of ivy or submerged and aerial leaves of some water plants. These different forms are **dimorphic**.

**diploid:** Possessing two sets of chromosomes per nucleus, cell, or organism; the 2n number. One set is paternal, the other maternal. This chromosome complement characterizes a species or a sporophyte generation in plants with alternation of generations.

**direct embryogenesis:** Embryoid formation directly on the surface of zygotic or somatic embryos or on seedling plant tissues in culture, without an intervening callus phase.

**direct organogenesis:** Organ formation directly on the surface of relatively large intact explants, without an intervening callus phase.

**disaccharide:** A sugar composed of two monosaccharide units joined by a glycosidic linkage; as sucrose, maltose, lactose, etc.

**disease:** A physiological or anatomical plant disorder or disturbance of biotic (pathological) or abiotic (environmental) origin that interferes with its structure, function or value; as some plants suffer from nutritional deficiency **diseases** in culture.

**disease-free:** An apparently healthy plant that shows no visible symptoms of disease and repeatedly tests negatively for specified pathogen(s). Such plants can be certified disease-free or specific pathogen tested (SPT). Disease-free plants generally form the basis for micropropagation or other forms of bulking-up vegetatively propagated plants.

**disease resistance:** The ability to resist disease or the agent of disease (the vector) and so remain healthy. Resistance or tolerance to disease is a topic of intensive interest in plant tissue culture work. Screening or selection at the cellular level is followed by plant regeneration and screening of these and their progeny.

**disinfect:** To free from infection by destroying disease-causing microorganisms, usually by means of a physical or chemical agent (**disinfectant**) which kills infectious organisms. The process is **disinfection**.

**dispense:** To give, deal or portion out; as nutrient medium is portioned into glassware for plant tissue culture.

**dissect:** To cut, expose, separate or divide the parts of a plant or animal into sections.

**dissecting microscope:** A relatively low power microscope used to facilitate dissection, examination or excision of small plant or animal parts. These microscopes usually magnify at least 50 times.

**dissociate:** **1.** To disconnect, come apart or disunite; as friable callus disperses into single cells and small aggregates when placed in suspension culture. **2.** Electrolytes in solution dissociate when ionized.

**dissolve:** To cause to pass into solution; as chemicals are added to (**dissolved**) in water in the preparation of nutrient solutions.

**distal:** Located away from the point of attachment of an organ.

**distilled water:** Water that has been converted to steam and the vapour recondensed. This process removes dissolved materials, particulates and microorganisms. The process may be repeated (double distilled) for added purity, often in a glass apparatus (glass distilled) to minimize metal recontamination. Water of this purity is commonly used to make nutrient media for plant tissue culture.

**diurnal:** During the day or occurring daily; at least once every 24 hours.

**DNA:** See **deoxyribonucleic acid**.

**dominant:** **1.** Designating a gene (dominant gene) that, appearing in a hybrid offspring, expresses its full phenotypic effect, excluding the expression of its allele. **2.** Describing a character (dominant character) possessed by one parent which in a hybrid masks the expression of a different (recessive) character derived from the other parent. **3.** In ecology, describing an individual or species that to a large extent controls the conditions for existence of its associates.

**donor plant (mother plant):** The source plant used for propagation, whether an explant, graft or cutting. Source plants used for micropropagation are usually pathogen-tested (disease-free).

**dormancy:** A period in which growth or functional (physiological) activity in seeds, bulbs, buds or other plant organs slows or stops (resting stage). Growth resumes only if specific environmental or physiological requirements are fulfilled.

**dormin:** See **abscisic acid**.

**dosimeter:** An instrument that measures the total energy radiation absorbed per unit time (**dose**).

**dry ice:** Frozen (solid) carbon dioxide ($CO_2$). It is commonly used as a refrigerant.

**dry weight:** The moisture-free weight of tissue obtained by drying at high (oven-drying) or low (freeze-drying) temperatures for an interval sufficient to remove all water.

**dual culture:** A culture system that includes plant tissue and one organism (such as a nematode species) or microorganism (such as a fungus). Dual cultures are used to study host-parasite interactions or in the production of axenic cultures for a variety of purposes. Usually the microorganism selected is an obligate parasite.

# E

**edamin:** See **casein hydrolysate.**

**EDTA:** See **ethylenediaminetetraacetic acid.**

**efflorescence:** See **anthesis.**

**egg:** See **ovum.**

**egg aparatus:** The structure containing the embryo sac, within which is the egg cell (ovum).

**elaioplast or lipidoplast:** An oil-storing plant cell plastid. A specialized leucoplast.

**electrolyte:** A substance which dissociates in aqueous solution into ions allowing electrical current to be carried by ion movement.

**electromagnetic radiation:** This region of the electromagnetic spectrum includes ultra violet (u.v.), and x and gamma radiation (x- and $\gamma$-rays). These are wavelengths used to produce genetically mutant cells or organisms and in the case of u.v., disinfection and sterilization, in plant tissue culture.

**electron microscope (EM):** A microscope employing an electron beam to image objects. Electrons have very short wave-lengths compared to light, so have a much greater resolving power. Such instruments are capable of magnifications 200,000 times or higher. Variations of this technology include the transmission electron microscope (TEM) and the scanning electron microscope (SEM). In the TEM the electron beam passes through the object, which must be very thin, and is scattered in a characteristic way and imaged through a magnetic or electrostatic focusing system on a fluorescent screen. In the SEM a thick sample is used. A three dimentional screen image is acquired through focusing secondary electrons emitted from the sample surface.

**element:** A substance composed of atoms of the same atomic number that are chemically uncombined. These **elements** cannot be decomposed by ordinary chemical means.

**ELISA:** See **enzyme-linked immunosorbent assay.**

**EM:** See **electron microscope.**

$\mu Em^{-2}s^{-1}$: See **$\mu molm^{-2}s^{-1}$.**

**embryo:** **1.** A rudimentary plant (sporophyte) formed from the fertilized egg cell in a seed (zygotic embryo) or through parthenogenesis. A common explant source (intact or a portion thereof) for tissue culture (embryo culture) at various stages of development (pro-embryo, globular, heart-shaped, torpedo-shaped, cotyledonary). The seed plant embryo consists of cotyledon(s), epicotyl, hypocotyl and radicle. Its development is termed **embryogenesis** or **embryogeny**. **2.** A somatic embryo is morphologically similar to a zygotic embryo but is initiated from a somatic cell(s). Nonzygotic **embryos** (embryoids) develop into plantlets in vitro through developmental processes similar to those of zygotic embryos.

**embryo culture:** **1.** Denotes a culture in which the explant was an embryo. **Embryo cultures** have been used to obtain viable offspring from seeds with a tendency for embryo abortion or when viable seed are limited in number. **2.** Cultures in which embryos are induced to form (embryogenesis) whether in suspension or on a variety of explants or cultures on solidified media. More correctly, these nonzygotic or somatic embryos are termed embryoids.

**embryo sac:** A structure produced by the megaspore mother cell in the ovules of angiosperms. At the micropylar end is the egg cell, which may be associated with 0–2 synergids. At the chalazal end are 0-several antipodal nuclei. In the middle of the embryo sac are one or more polar nuclei. There are at least ten types of embryo sacs among the angiosperms.

**embryogenesis or embryogeny:** The process of embryo or embryoid formation, whether by sexual (zygotic) or asexual means. In asexual embryogenesis embryoids arise directly from the explant or on intermediary callus tissue. In some cases they arise from individual cells (somatic cell embryogenesis). The hazards of asexual embryogenesis involve the possibility of geneotypic or phenotypic variation in the propagules.

**embryoid:** See **somatic embryo**.

**empirical:** Relating to or based upon practical experience, trial and error, direct observation or observation alone without benefit of scientific method, knowledge or theory.

**$\mu Em^{-2}s^{-1}$:** See **microeinstein**.

**emulsify:** To make an **emulsion**; suspend fatty or resinous material in a liquid, usually through agitation.

**enation:** Outgrowth on a plant surface.

**endobiotic:** Living inside another organism; as do some bacteria (endogenous bacteria) and other **endobionts**.

**endogenous:** Arising from internal tissues or processes; as hormones synthesized by plant tissues. The contrasting process is exogenous or external; as substances included in nutrient media.

**endogenous inhibitor:** A substance produced by an organism capable of inhibiting its own growth or some specific physiological function or that of another organism exposed to it even at low concentrations; as abscisic acid (ABA).

**endomitosis:** The process of chromosome doubling within the intact nuclear membrane, and not followed by chromosome movement or cytokinesis. The result is a diploid, tetraploid, octaploid or other polyploid homozygote. This is a common consequence of callus culture.

**endopolyploidy:** Polyploidy resulting from nuclear division (karyokinesis) without subsequent cytoplasmic division (cytokinesis).

**endoreduplication:** The process of chromosome reproduction during interphase; as 4-chromatid chromosomes (diplochromosomes) may be seen during interphase.

**endosperm:** A nutritive tissue contained in the angiosperm ovule, commonly triploid, developed in the embryo sac by the union of a male sperm nucleus and one or more polar nuclei. This nucleus divides to form the endosperm. Endosperm is a complex substance containing amino acids, sugars and hormones. It persists in some seeds and is used by the embryo and seedling at germination. Coconut milk is derived from endosperm tissue and is a useful addendum in some tissue culture media.

**environment:** The conditions surrounding an organism including biotic (living organisms both harmless and harmful) and abiotic (climatalogical, edaphic, etc.) factors which in concert with the genotype affect its growth and welfare (phenotype).

**environmental chamber:** A controlled environment cabinet (incubator) in which temperature, light quality, intensity and duration; and preferably also the relative humidity and airflow are controlled.

**environmental variance:** Phenotypic variance resulting from the environmental conditions to which an organism is exposed.

**enzyme:** Any one of a number of specialized proteins produced in living cells, that speed the rate of specific chemical reactions, even at very low concentrations (organic catalyst), but is not used up in the reaction.

**enzyme-linked immunosorbent assay (ELISA):** A sensitive serological test used for detection and quantification of viruses, proteins and small molecules such as hormones. Performed on a microtiter plate, many samples can be tested at the same time both rapidly and economically. The most widely used form of ELISA testing uses the double antibody sandwich which involves the addition of a specific antibody to the test plate, where it adsorbs, followed by addition of the test sample. Specific particles in the sample are immobilized on the antibody film. Subsequently, enzyme-labeled antibody is added and becomes immobilized on the sample particles. Test particles are then quantified by the addition of enzyme substrate, through the colorometric or fluorometric detection of the reaction product.

**epicotyl:** The upper part of the embryo or seedling axis above the point of the cotyledon(s) attachment. It develops into the stem and leaves.

**epidermis:** The outermost (superficial) layer of cells of leaves, young roots and stems of plants. It is living, parenchymatous and primary in origin and develops from the protoderm. The epidermis, like skin in mammals, prevents physical injury and excessive water loss. A cuticle usually covers the epidermis of aerial plant parts and reinforces its protective function. Some epidermal cells are modified; as the guard cells and hairs. The epidermis is replaced by the periderm during secondary growth of stems and roots.

**epigenetic:** **1.** Describes environmentally induced variations in the phenotype perpetuated by cloning but not involving permanent changes in the genotype. **2.** Resulting from the interaction of genetic factors that occurs during the normal development of an individual plant. These phenotypic changes, whether resulting from internal (ontogenetic, hormonal) or external (temperature, light, etc.) causes, once induced persist in the absence of the inducing agent. The mechanisms of induction and the course of changes in gene expression are unclear; apparently involving gene to enzyme information transfer (tRNA systems). Epigenetic phenomena include habituation, determination (topophysis) and juvenile or adult phase changes. The study of the causes of developmental variation is **epigenetics**.

**epinasty:** An abnormal downward-pointing growth of branches or petioles due to more rapid growth of the upper side. This habit may result fom either nutritional deficiencies or hormonal irregularities and should not be confused with wilting, as **epinastic** tissues are turgid.

**epistomatic:** Refers to leaves having predominantly adaxial stomata; abaxial stomata are fewer than 10% of adaxial stomatal numbers.

**equatorial plate:** In metaphase (mitosis or meiosis) this is the plane of separation between the two sets of chromosomes. In mitosis, the centromeres lie on the plate but in meiosis the contromeres lie on the spindle or opposite sides of the plate and equidistant from it.

**equimolar:** The same amount of solute per litre of solution, the same **molar** concentration.

**Erlenmeyer flask:** A wide-necked variety of conical, flat-bottomed flask commonly used for medium preparation. Smaller versions are used as culture containers. Named after E. Erlenmeyer (1825–1909).

**essential requirement:** A nutrient mandatory for growth and development, reproduction and good health. In vitro cultures require inorganic salts, including all of the elements presently viewed as necessary to plant metabolism, organic factors (amino acids, vitamins), usually hormones (auxins, cytokinins and often gibberellins) as well as a carbon source, usually as sucrose or glucose.

**established culture:** **1.** Achievement of an aseptic viable explant, synonymous with Stage I culture (culture establishment) or transplant, synonymous with Stage IV culture (establishment in soil). **2.** A suspension culture subjected to several passages, adjusted so that cell number per unit time is constant from subculture to subculture.

**ethanol or ethyl alcohol: ($C_2H_6O$ mw 46.07):** Used to disinfect plant tissues, utensils, glassware and working surfaces in plant tissue culture manipulations. Typically the concentration used is 70% (v/v) for disinfecting and 95% (v/v) when flaming tools. It is sometimes used to dissolve water-insoluble additions to culture media.

**ethephon or (2-chloroethyl)phosphonic acid ($ClC_2PO_3H_6$ mw 144.50):** A synthetic compound that spontaneously degrades to release ethylene. Often used as a means to treat cultured cells (or unripened fruit) with ethylene.

**ethyl group:** A two carbon ($CH_3CH_2$-) chain component of many organic molecules. For example ethanol is $CH_3CH_2OH$.

**ethylene or ethene ($C_2H_4$ mw 28.25):** A simple hydrocarbon, colorless, inflammable gas with plant growth regulatory action. Its production is stimulated by auxins, wounding and stress. It is involved in fruit ripening; promotion of flowering; root induction; stimulation of seed germination; promotion of leaf abscission and inhibition of auxin transport. It is produced by some plant tissues in culture and may repress embryogeny in some plants and cause epinasty in others.

**ethylene glycol:** See **polyethylene glycol**.

**ethylenediaminetetraacetic acid, disodium salt ($Na_2EDTA$, $C_{10}H_{14}N_2O_8Na_2.2H_2O$ mw 372.25):** A chelating agent which reversibly binds positive ions such as iron, magnesium, etc. It is commonly added to plant tissue culture media to keep iron and other salts available by releasing them slowly into the medium as required.

**ethylmethanesulfonate (EMS, $C_3H_8SO_3$ mw 124.14):** A very potent chemical mutagen frequently used in mutagenic studies. It acts by adding ethyl groups to guanine, subsequently causing base pairing errors as it binds to adenine.

**etiolation:** The condition seen when plants grow under low levels or absence of light, causing the stem to become spindly and elongated, petioles extended, leaves poorly developed, internodes very long and levels of chlorophyll low. Light reverses these effects.

**etiology or aetiology:** The underlying cause or reason, the predisposing causes of either abiotically or biotically induced events, especially disease.

**eukaryotic or eucaryotic:** Having a true nucleus in cells or organisms. The nucleus is separated from the cytoplasm by a nuclear membrane and the genetic material is located on chromosomes within the nucleus. Nuclear division is through mitosis or meiosis, unlike in prokaryotic cells where it is amitotic.

**euploid:** A cell with the exact multiple of the monoploid chromosome number in the nucleus; as haploid, diploid, tetraploid, etc. The condition is **euploidy**.

**evergreen:** Vascular plants that produce and shed leaves throughout the year. Their branches are never bare like those of deciduous trees.

**ex vitro:** (Latin: "from glass"). Organisms removed from culture and transplanted, generally to soil or potting mixture.

**excise:** To cut, sever or otherwise remove or extract an organ or a segment of tissue from a plant or plant part; as the surgical removal of shoot tips. The process is **excision**.

**exogenous:** Originating from without or resulting from an external cause. In this category are stresses imposed on cultured plants by the environment, especially the culture medium; as through the influence of exogenous hormones. The antonym is endogenous.

**explant:** The excised plant portion used to initiate a tissue culture. The process of dissection and removal to culture of these small organs or tissue sections is **explantation**. Explant choice, the timing of excision and pretreatment are important determinants of culture success.

**explant donor:** The source plant or mother plant from which the explant used to initiate a culture is taken.

**exponential phase:** A phase in culture in which cells undergo their maximum rate of cell division. It follows the lag phase and preceeds the linear growth phase in most batch propagated suspension cultures.

**exude:** To discharge slowly, leak liquid material (**exudate**) sometimes through pores or cuts but often by diffusion into the medium. This process of **exudation** is associated with a lethal browning of explants in some woody plant species. These in vitro **exudates** have not been chemically characterized but are often referred to as tanins or oxidized polyphenols.

# F

**$F_1$:** Offspring (hybrid); the first filial generation following a cross of the parental (homozygous) generation ($P_1$). **$F_2$, $F_3$** are the second and third generations from crosses between the **$F_1$, $F_2$** generations respectively.

**factor:** **1.** A condition, influence or constituent to be taken into account. **2.** A genetic character determinant (gene).

**facultative:** Flexible. The capacity to live under more than one set of environmental conditions or to use more than one energy source.

**falcate or falciform:** Curved or sickle-shaped; as in the shape of some leaves and leaf hairs.

**Farmer's fixative:** **1.** 3 Anhydrous ethanol:1 glacial acetic acid; a fixing and dehydrating agent used in histology. **2.** Used in conjunction with feulgen stain and carbolfuchsin stain for chromosome analysis.

**Fe:** The chemical symbol for the element **iron**.

**fermentation:** The anaerobic energy-yielding breakdown (decomposition) of complex organic substances, especially carbohydrates like glucose, by microorganisms. It is often misused by tissue culturists to describe large scale aerobic plant cell culture for secondary product synthesis.

**ferric ethylenediamine tetraacetate, sodium salt (NaFe.EDTA $C_{10}H_{12}FeN_2NaO_8$ mw 367.07):** A microelement **iron** and **sodium** salt added to some plant tissue culture media. It may be substituted for $FeSO_4$ + $Na_2$.EDTA however, if substituted on an equimolar basis, while the same amount of Fe will still be available, half the Na will be lacking and may have to be added in the form of another salt.

**fertilization:** The significant feature of sexual reproduction. Refers to gametic fusion, especially the fusion of their nuclei doubling the chromosome number and forming a zygote. In double fertilization, which occurs in many angiosperms, one male nucleus fuses with the egg nucleus as a second male nucleus fuses with two female polar nuclei to produce the endosperm, a triploid nutritive tissue. Test tube fertilization involves pollination and fertilization in vitro.

**Feulgen's test:** A test used to detect nuclear DNA, especially during cell division. The procedure involves tissue incubation in dilute acid to hydrolyze DNA, exposing the aldehyde groups of deoxyribose. This is followed by soaking in Schiff's reagent, which colors the DNA a deep magenta.

**fibre:** An elongated, thick-walled, often lignified cell (sclerenchyma) present in various plant tissues and providing mechanical support. These occur in the xylem (xylary fibres) or outside the xylem (extraxylary fibres).

**Ficoll:** The brand name for an inert, synthetic, highly soluble polymer with use as an osmotic agent sometimes used for suspending protoplasts.

**field test:** An evaluative test whereby the field performance of experimental plants is assessed in comparison to controls.

**filament:** A slender threadlike structure; as the stalk of a stamen supporting an anther (in flowers).

**filial generation:** See $\mathbf{F_1}$.

**filter paper:** A porous paper used for a multiplicity of scientific purposes which include **filtration** and tissue support during culture.

**filter sterilize:** To sterilize by passing a solution through a porous material capable of separating out suspended microbes or their spores; as heat-labile components of nutrient media are sterilized.

**filtration:** **1.** A process of separating solids from liquids using a porous material that retains the solid and allows passage of the liquid only or solid of a size smaller than the pore size (**filtrate**). **2.** The process of washing a cell suspension through decreasing pore size meshes (**filters**) that remove cell aggregates to obtain a **filtrate** of single cells that can be utilized as plating inocula.

**fixative:** A compound that stabilizes, sets or **fixes** other compounds or structures securely so that their structural integrity is retained; as the process of **fixation** utilizes chemical agents to permanently prepare cells or tissues for microscopy.

**flame sterilize:** To sterilize tools or instruments usually by heating them in a flame until they glow. This procedure is used in conjunction with ethanol (70%) emersion or ethanol (95%) dip. The most common means of sterilizing forceps and scalpels used in sterile tissue culture manipulations.

**floccule:** An aggregation (coalescence) of microorganisms or colloidal particles floating in or on a liquid. **Flocculation** is seen in some contaminated liquid media appearing as a cloud.

**florescence:** Refers to anthesis or flowering time, the state of being in bloom.

**floricane:** The flowering and fruiting stem (2nd year) of biennially producing plants such as *Rubus* species. The first year canes are vegetative and are called primocanes.

**florigen:** An hypothetical hormone-like substance implicated in floral initiation in many plant species. This substance is graft-transmissable and is said to explain transmission of the photoperiodic flowering stimulus from leaves to growing points, where it induces floral initiation.

**flower:** The angiosperm reproductive structure; may include calyx, corolla, androecium (stamens) and gynecium (carpels). These are common explant sources for plant tissue culture.

**fluid drilling:** A mechanical procedure for planting seed; pre-germinated seeds are suspended in a gel and sowed through a **fluid drill** seeder. This technology is potentially adaptive for sowing artificial seed (somatic embryos or embryoids).

**fluorescein diacetate (FDA mw 416.39) stain:** Living cells stained with FDA fluoresce in the presence of ultra violet light. This stain is used to assess cell viability (percent viability) of cultures.

**fluorescein isothiocyanate (mw 389.39):** A compound used to stain plant protoplasts; as in the identification of fusion products.

**fluorescence:** The absorption of light of a specific wavelength followed by emission of light of a longer wavelength.

**foam plug:** A closure made of reusable, autoclavable, spongy foam.

**fog:** Fine particles of liquid suspended in the air; as with water in a **fog chamber** used for acclimatizing recent ex vitro transplants.

**foliar:** Pertaining to leaves.

**folic acid ($C_{19}H_{19}N_7O_6$ mw 441.40):** A member of the vitamin B group (Bc). It is also known as vitamin M. It is present in green leaves and has some coenzyme activities. It is occasionally added to plant tissue culture media.

**foot candle (fc):** An obsolete measurement unit of light intensity (ca. 10 lux). The recommended light intensity units are $\mu molm^{-2}s^{-1}$, $jm^{-2}$ or $wm^{-2}$.

**forceps:** A hand-held, pincer-like instrument designed to grasp, hold or pull objects in delicate manipulative work; as in plant tissue culture manipulations.

**formaldehyde:** See **formalin**.

**formalin:** Aqueous formaldehyde ($CH_2O$), 37–50% by weight in water (w/v), 10–15% methanol; with preservative and antiseptic properties.

**formazan:** Colorless when dissolved in water, the chemical 2,3,5-triphenyl tetrazolium chloride is reduced to the red colored chemical triphenyl formazan on contact with living, respiring tissue. The amount of formazan formed is used as a measure of seed viability, as it reflects oxidative metabolism.

**formula:** A prescribed or set form, method or recipe. Devising the original formula is **formulation**.

**formula weight (FW):** The gram molecular weight of a chemical compound.

**fortify:** To enrich or add strengthening components; as medium can be **fortified** by the addition of a beneficial ingredient.

**freeze-dry or lyophilize:** To dry in a frozen state under vacuum; as tissues are **freeze-dried** to obtain a dry weight or to preserve them for analysis.

**freeze preservation:** See **cryopreservation**.

**fresh weight or wet weight:** The weight of a plant or plant part including the water content.

**friable:** Readily crumbling or fragmenting; as is the preferred callus type to initiate cell suspension cultures, as it is easily dissected with a spatula and readily dispersed into single cells or small clumps of cells in solution.

**fructose or fruit sugar or levulose ($C_6H_{12}O_6$ mw 180.16):** A hexose sugar commonly found in plants. Combined with glucose, it forms sucrose. Fructose is an occasional carbohydrate source in nutrient solutions used for plant tissue culture, for energy or as an osmoticum.

**fruit:** The ripened ovary of a flower and any associated accessory parts.

**fuchsin:** See **carbolfuchsin**.

**fungus, pl. fungi:** Any of a large group of sometimes multicellular, multinucleate, achlorophyllous, eukaryotic organisms with chitin-containing cell walls, surviving parasitically or saprophytically (using an absorptive mode of nutrition) and reproducing by hyphal growth or spores. Fungi are common contaminants of plant tissue cultures.

**funiculus or funicle:** The stalk anchoring and supplying the vascular supply to the ovule, and later the seed, from the ovary wall or placenta in angiosperms.

***N*-(2-furanylmethyl)-1*H*-purin-6-amine or 6-furfurylaminopurine or furfuryladenine or kinetin (K, sometimes Kn, $C_{10}H_9N_5O$ mw 215.22):** A growth hormone of the cytokinin-type isolated from animal and plant DNA, and capable of promoting plant cell division. It is often used in plant tissue culture media. It is soluble in dilute HCl (ca. 1 M). The first cytokinin discovered was extracted from herring sperm DNA by Miller et al., 1955.

**fusion:** The merging of gametes, cells, protoplasts or other structures together to form zygotes, hybrids, cybrids or other fusion products.

**fusogen:** A fusion-inducing agent used for protoplast agglutination in somatic hybridization studies; as is polyethylene glycol.

# G

**$GA_3$:** See **gibberellic acid**.

**galactose ($C_6H_{12}O_6$ mw 180.16):** A hexose sugar found in plants. A component of the disaccharide lactose and many plant polysaccharides. Galactose is an infrequently used carbohydrate source in nutrient solutions used in plant tissue culture.

**galacturonic acid:** A hexose sugar acid derived from galactose.

**gall:** A discrete, abnormal plant outgrowth or swelling induced by one of a number of pathogenic organisms.

**Gamborg, O.L. (1966):** Formulated PRL-4, predecessor to the B5 medium formulation; a medium commonly used in plant tissue culture for rapid cell proliferation.

**Gamborg, O.L., R.A. Miller and K. Ojima (1968):** Formulated B5, a medium commonly used in plant tissue culture for rapid cell proliferation.

**gamete or germ cell:** Reproductive cell with half the somatic cell chromosome number (n). In fertilization two **gametes** fuse to form a diploid (2n) zygote that develops into a new individual.

**gametoclone:** A plant regenerated from a tissue culture originating from gametic tissue.

**gametophyte generation:** The haploid (n), gamete-producing phase of the life cycle in plants with alternating generations.

**gamma radiation ($\gamma$):** Short wavelength electromagnetic radiation emitted from the radioactive decay of atomic nuclei.

**Gautheret, R.J.:** A French scientist who obtained the first callus cultures from excised cambial tissues of woody species (1934). By including in his medium White's B vitamins and the auxin IAA he later established the first callus cultures capable of continuous proliferation (1939). He also discovered habituation.

**gel:** A mixture of thick or firm consistency containing liquid trapped in a solid component; as **gelling** agents such as agar solidify liquid nutrient media. The process is **gelation**.

**gelatin:** A glutinous, proteinaceous material produced by boiling, which partially hydrolyzes the collagen, of animal connective tissues. It is sometimes used to **gel** or solidify nutrient solutions for plant tissue culture. Other **gelling** agents, specifically agars, are usually preferred.

**Gelrite:** The brand name of a synthetic (*Pseudomonas*-derived) refined polysaccharide used as a gelling agent and agar substitute.

**gene:** A self-duplicating segment of a DNA molecule (chromosome) made up of several hundred nucleotides, specifying number, kind and arrangement of the amino acids in a polypeptide chain or RNA molecule. The smallest functioning unit (hereditary factor or unit) of the chromosomal **genetic** material determining one or more of the hereditary characteristics of an organism and capable of mutation.

**generate:** To propagate or (mass) proliferate. The process is **generation** or **regeneration**.

**generation time:** The time between successive generations of cells or organisms within a population.

**generative:** Refers to a somatic cell or tissue.

**genetic engineering or recombinant DNA technology:** Technology involving man-made changes in the genetic constitution of cells (apart from selective breeding). This technology usually employs a vector (such as the Ti plasmid of *Agrobacterium tumefaciens*) for transferring useful genetic information from a donor organism into a cell or organism that does not possess it. Genetic engineering has many potential uses.

**genetic expression:** The expression of hereditary information; as seen in the phenotype (anatomical, physiological, biochemical character) of an organism.

**genetic information:** The hereditary information contained in a nucleotide base sequence in chromosomal DNA or RNA.

**genetic selection:** Selection of genes, cells (cell selection), clones, etc., by man within populations or between populations or species. The usual purpose is to alter a specific phenotypic character. Such selection usually results in differential success rates of the various genotypes, reflecting many variables including selection pressure and genetic variability in populations.

**genetic transformation:** The transfer of extracellular DNA (genetic information) among and between species; as, for example, with the use of bacterial or viral vectors.

**genetic variation:** Differences in individuals derived from the same genotype in distinction to differences caused by the environment. **Genetic variance** denotes the proportion of phenotypic variance caused by differences in the genetic make-up of an individual.

**genetics:** The science of hereditary development and variation between organisms.

**genome:** The basic haploid chromosome set of an individual; the sum of its **genes**.

**genotoxic:** Carcinogenic; toxic to the chromosomes.

**genotype:** The genetic constitution of an organism. Its genetic information (complement of **genes**), and so its hereditary potential, in contrast to its phenotype. The genotype determines the range of phenotypic responses an organism possesses (the reaction norm) in relation to its environment.

**geotaxis:** Plant orientation with respect to gravity.

**geotropism:** Plant growth movement in response to gravity. The orientation of some plant parts is parallel to the lines of gravity while main stems are negatively **geotropic** and main roots are positively geotropic. This tropism is easily illustrated in plant culture systems (although in some cases roots grow up and shoots down in culture). The orientation mechanism is thought to involve the distribution of auxins.

**germ:** **1.** A disease-causing microorganism. **2.** The common name for a plant embryo.

**germ cell:** A reproductive cell such as a gamete or its primordial cell (gametophyte) which is quite distinct from a somatic cell.

**germicide:** Any chemical agent that kills pathogenic microorganisms.

**germination:** The sprouting and development of vegetative growth of an embryo, spore, seed or other reproductive body forming a new individual plant.

**germplasm or germ plasm: 1.** The reproductive body tissues distinct from somatic (nonreproductive tissues). The genetic material, basis of heredity of an organism, passed on through previous generations. **2.** An individual representing a type species or culture that may be held in a repository for agronomic, historic (or other) reasons.

**gibberellic acid (gibberellin $A_3$ or $GA_3$, $C_{19}H_{22}O_6$ mw 346.37):** One of the **gibberellins**, a group of growth hormones promoting cell division and elongation. The first of the group to be isolated and the most widely used gibberellin in plant tissue culture. Isolated from the fungal pathogen *Gibberella fujikuroi*. It dissolves in base (ca. 1M KOH or NaOH).

**gibberellin:** One of a group of more than 50 terpenoids which are plant growth hormones (phytohormones) and growth regulators exhibiting physiological activity similar to **$GA_3$**; with a gibbane ring skeleton. Its biological activity includes promotion of cell enlargement and stem elongation (by increasing cell wall expansibility); flowering; parthenocarpic fruit growth and dormancy-breaking in seeds and plant organs. Auxins interact with **gibberellins** to control some growth actions, abscissic acid modifies some others. Gibberellins are used in plant tissue culture media to stimulate new growth and promote shoot formation and elongation. Stock solutions are prepared by dissolving the hormone in base (ca. 1M KOH or NaOH), then making up to volume with water.

**gland:** A unicellular or multicellular structure with a secretory function, discharging chemical substances to the plant surface via glandular hairs, hydathodes or nectaries or internally through resin canals.

**glabrous:** Lacking hairs or projections.

**glaucus:** A surface with a waxy, white coating that can be rubbed off.

**glucose, dextrose or grape sugar ($C_6H_{12}O_6$ mw 180.15):** A photosynthetically formed monosaccharide (hexose) sugar widely found in plants in many polysaccharides (starch, cellulose). It is a commonly used carbohydrate source in plant tissue culture media, for energy and as an osmoticum.

**glucoside:** See **glycoside**.

**glutamic acid or glutamate (Glu, $C_5H_9NO_4$ mw 147.13):** An amino acid involved in nitrogen metabolism. It is added to some plant tissue culture media as a source of reduced nitrogen.

**glutamine (Gln, $C_5H_{10}N_2O_3$ mw 146.15):** An amino acid involved in purine biosynthesis, occasionally added to plant tissue culture media. It may replace ammonium ($NH_4^+$) ions as the nitrogen source. It is of key importance in pollen growth in vitro.

**glutathione ($C_{10}H_{17}N_3O_6S$ mw 307.33):** A major low molecular weight thiol compound found in plants. It serves as a hydrogen acceptor in many reactions. It is an occasional additive in plant tissue culture media.

**glycerol or glycerin(e) ($C_3H_8O_3$ mw 92.09):** A thick, clear, trihydroxy alcohol miscible with water or alcohol and a by-product of fat hydrolysis. It is commonly employed as a cryoprotectant and as a temporary slide specimen-mounting medium.

**glycerophosphatide:** See **phospholipid**.

**glycine or aminoacetic acid (Gly, $C_2H_5NO_2$ mw 75.07):** The simplest amino acid, synthesized and broken down through several pathways. It is a component of porphyrins and is essential in purine synthesis. It is a common addendum to nutrient solutions for plant tissue culture.

**glycoside:** An ether-type derivative of sugar and hydroxy compounds occurring widely in plants. **Glycosides** derived from glucose sugar are glucosides.

**glycosidic bond or glycosidic linkage:** The basic bond linking monosaccharide units of disaccharides and polysaccharides.

**gnotobiotic:** The growth of an organism alone (sterile) or in the presence only of known organisms.

**graft:** To induce the union of a small part of a plant (scion) to a relatively large one (stock) that is usually separate on the same organism (**autograft**), another organism (**homograft**) or even another species (**heterograft**). In this way the desirable traits of both graft partners are obtained. Many cultivated varieties of fruit trees and ornamentals, such as roses, are propagated by **grafting**.

**graft inoculation test:** A viral indicator test in which a suspected viral carrier is grafted to an indicator plant. If symptoms appear the viral assay is positive.

**gram (g or gm):** The metric unit of mass and weight equal to 1/1,000 kilogram or 1 $cm^3$ water at its maximum density (4°C).

**Gram's iodine:** See **iodine**.

**Gram's stain:** A staining technique used to distinguish between two major bacterial groups based on stain retention by their cell walls. Bacteria are heat-fixed, stained with crystal violet, a basic dye, then with iodine solution. This is followed by an alcohol or acetone rinse. Gram-positive bacteria are stained bright purple. Gram-negative bacteria are decolorized, so safranin is used to stain them.

**grana, sing. granum:** Stacks of circular thylakoids, composed of lamellae in higher green plant chloroplasts, containing pigments and other essential constituents of photosynthetic light reactions. This thylakoid arrangement is thought to increase the efficiency by which photosynthetic light energy is trapped and converted to chemical energy.

**grape sugar:** See **glucose**.

**Gro-lux:** The brand name for a type of wide spectrum fluorescent lamp.

**ground meristem:** One of three primary meristematic tissues derived from the apical meristem and differentiating into tissues other than epidermal and vascular.

**growth:** **1.** An irreversible increase in cell size and/or number, resulting from cell division or enlargement and usually accompanied by differentiation. **2.** Maturation; differentiation and morphogenesis.

**growth cabinet:** See **cabinet**.

**growth inhibitor:** Any substance (its own or that of another organism) that inhibits the growth of an organism. The inhibitory effect can range from mild inhibition (growth retardation) to severe inhibition or death (toxic reaction). Two hormones that may act as inhibitors are ethylene and abscisic acid (ABA). The concentration of the substance and the length of exposure to it (dose) determines its effects, as do other factors such as the relative susceptibility of the organisms exposed to it.

**growth rate:** Increase in size per unit of time.

**growth regulator:** Any organic substance, other than a nutrient, which in small amounts induces a change in or otherwise influences plant growth and development. This body of substances includes plant hormones, substances derived from non-plant sources and laboratory-synthesized substances.

**growth substance:** Any organic substance, other than a nutrient, which in small amounts induces a physiological, developmental, timing, or size change response in plant cells, tissues or organs. This body of substances includes plant hormones, growth regulators and many herbicides.

**guanine or 2-aminohypoxanthine ($C_5H_5N_5O$ mw 151.27):** One of two purine nitrogenous bases occurring in the nucleotides of nucleic acids, which are integral to the formulation of the genetic code. Guanine differs from adenine in that the $N^6$ amine is replaced by an oxygen atom.

**guard cell:** A specialized kidney-shaped epidermal cell, one on each side of a stomatal pore forming a stoma, causing stomatal opening and closing based on changes in turgor brought about by metabolic changes in the plant. These cells regulate gas and water vapor exchange from the leaves of plants.

**Guha, S. and S.C. Maheshwari (1966):** Were among the first to obtain haploid (n) plantlets from anther culture.

**guttate:** To exude water, salts or other materials through hydathodes of plant leaves. The **guttation** occurs when the relative humidity is high and transpiration is prevented or when transpiration rates are maximum, but insufficient to relieve root pressure.

**gymnosperm:** A member of a group of vascular plants (class Gymnospermae) of about 700 species, whose seeds are not enclosed within an ovary (naked seeds) and in which double fertilization does not occur so the seeds possess no endosperm. For example, conifers.

**gynoecium:** A collective term for the carpels of a flower; the female reproducive organs.

**gynogenesis or pseudogamy:** A form of parthenogenesis in which ovum development is initiated by the male gamete, although nuclear fusion is not involved, or the male nucleus is eliminated after fertilization.

# H

**H:** The chemical symbol for the element **hydrogen**.

**Haberlandt, G.:** An inspirational pioneer botanist who was the first to attempt to culture plant cells at the turn of the 20th century. He predicted the existence of hormones that would stimulate cell division (auxins) and suggested the possibility of exploiting the totipotentiality of plant cells.

**habit:** The growth form of a plant; the characteristic arrangement of its organs and their shape, size, color and posture.

**habituated or anergized:** **1.** Describes a culture in which an exogenous requirement has been lost; as in the loss of an absolute nutritional or hormonal requirement. A change in morphology and the loss of organogenetic ability may also occur. **2.** Diminished response to a stimulus resultant from repetition. The process is **habituation**.

**haemocytometer:** See **hemocytometer**.

**hair:** A single or multicellular, sometimes absorptive (root hair) or secretory (glandular hair) and sometimes only a superficial outgrowth (covering hair) of the epidermal cells. The term trichome is often used but includes outgrowths from deeper within tissues. Distinguishing between hairs and trichomes is not always easy. However, trichomes usually have a vascular supply, while hairs do not.

**halophile or halophyte:** A "salt-loving organism"; one that is adapted to a high salt environment; as a salt marsh. These plants possess xeromorphic characters but are not able to withstand droughts, as xerophytes do.

**hanging droplet technique:** See **microdroplet array technique**.

**Hannig, E. (1904):** First to successfully culture (crucifer) embryos.

**haploid:** The reduced chromosome number or half the diploid (somatic) number (n); characteristic of the gametophyte generation. Having one set of chromosomes in the nucleus or in an organism. Haploid tissues or plants are useful in plant breeding.

**hardening or hardening off:** The process of making a plant tough, in preparation for increased stress. This may be accomplished by gradually reducing water or nutrients and increasing light or low temperature exposure.

**heat labile:** See **thermolabile**.

**heat pump:** A machine that extracts heat from a fluid or gas that is marginally above ambient temperature so that the temperature differential can be usefully employed. **Heat pumps** are used to heat (or cool) laboratories and greenhouses.

**heat therapy:** See **thermotherapy**.

**Heller, R. (1953):** Formulated a medium used in plant tissue culture. He also devised a paper bridge employed as a support structure and wick for tissues in liquid medium (the Heller bridge).

*Helminthosporium maydis* **race T toxin (HmT toxin):** A host-specific toxin produced by a fungal pathogen of corn, *H. maydis* race T. It is used for the in vitro selection of resistant corn callus.

**hemicellulase:** An enzyme from *Aspergillus niger*, available as a commercial preparation, that degrades hemicellulose to galactose.

**hemicellulose:** One of a group of plant cell wall polysaccharides similar to cellulose but with greater solubility and more easily hydrolyzed to simple sugars; as xylans, glactans, glucans and mannans. With pectin and lignin, **hemicelluloses** form the cell wall matrix.

**hemizygous or haplozygous:** Having a single-copy gene with no allele; as are all genes of haploids.

**hemocytometer or hemacytometer:** A ruled, calibrated, glass slide used with a microscope to count or examine cells. It was originally devised for counting red and white blood cells.

**HEPA filter:** An acronym for **high efficiency particulate air filter**. A filter capable of screening out particles larger than 0.3 μm. They are used in laminar air flow cabinets (hoods) for sterile transfer work.

**herb:** **1.** A nonwoody, perennial flowering plant whose aerial portion dies back at the end of each season. **2.** Any such plant valued for medicine, seasoning, scent, etc.

**herbicide:** A chemical agent that kills or inhibits the growth of herbaceous plants especially weeds.

**herbicide resistance:** Resistance or tolerance to **herbicides**. This is a field of intensive interest among those involved in plant tissue culture work. Screening or selection at the cellular level for putative resistance is tested by plant regeneration and followed by additional screening of these plants and their progeny.

**hereditary:** Refers to characteristics that are genetically transferable from one generation to the next.

**heteroauxin:** An obsolete term for the auxin 1*H*-indole-3-acetic acid (IAA).

**heteroblastic:** With both juvenile and adult (dimorphic) foliage, or a progressive increase in size and complexity in successively developed leaves, sometimes followed by a reverse in this tendency after flowering.

**heterogeneous:** With more than one type of cell or individual; as is callus, which is usually non-uniform in cell composition.

**heterograft or xerograft:** An interspecific graft. The likelihood of success of this graft is proportional to the degree of relatedness between the donor and recipient.

**heterokaryon or heterokaryocyte:** Refers to fused, unlike cells (multinucleate) with dissimilar nuclei. The occurrence is **heterokaryosis** and is common in fungi. The antonym is homokaryon.

**heterophylly:** The production of more than one leaf form on a plant species. In developmental heterophylly, juvenile leaves may differ from adult ones. In environmental heterophylly, differences may be environmentally induced; as on submerged and aerial leaves of some acquatic plants. In habitual heterophylly, variable leaf forms are habitually produced.

**heteroplasmon or heteroplast:** A cell with a mixture of two types of cytoplasm (cytoplasmic hybrid); cell with foreign organelles.

**heteroploid:** Cell (aneuploid), tissue or organism with lost or gained chromosomes; as one with other than a multiple of the monoploid number. The condition is **heteroploidy**.

**heterosis:** Hybrid vigor; the superiority of heterozygous genotypes to either parent.

**heterotroph:** An organism obtaining some or all of its food (complex nutrient molecules like glucose, amino acids, etc.) from external sources (formed by other organisms); requiring organic carbon. This term describes all animals, non-photosynthetic plants and some bacteria. In culture plants photosynthesize at a very low level, if at all, and are therefore **heterotrophic**. Ex vitro transplants must become fully photoautotrophic as soon as possible.

**heterozygote:** A cell or organism with both a dominant and a recessive allele, in the corresponding loci of a pair of homologous chromosomes. The condition is **heterozygous**.

**hexahydroxycyclohexane:** See **inositol**.

**hexitol:** A six carbon sugar alcohol, such as (meso-, myo-, or i-) inositol.

**hexose:** A monosaccharide containing six carbon atoms; as glucose, fructose or galactose. These are components of biologically important disaccharides and polysaccharides.

**high efficiency particulate air filter:** See **HEPA filter**.

**histine (His):** A precursor of histidine and one of the 20 common amino acids occurring in proteins. Infrequently added to plant tissue culture media.

**histogen theory:** The concept that specific tissues develop from plant meristematic tissue. The dermatogen, periblem and plerome (theoretically) give rise to the epidermis, cortex and primary tissues internal to the cortex, respectively. This has been replaced by the tunica-corpus theory.

**histology:** The study of tissue structure.

**homogeneous:** **1.** Of uniform consistency throughout. **2.** Made up of identical units.

**homograft or allograft:** A graft between individuals of like species.

**homokaryon or homokaryocyte:** Fused similar cells (multinucleate) with similar nuclei. The occurrence is **homokaryosis**. The antonym is heterokaryon or heterokaryocyte.

**homologous chromosomes:** Chromosomes with identical sets of loci; the two chromosomes of each of the pairs (one from each parent) normally found in somatic cells.

**homozygote:** A cell or organism with identical alleles (dominant or recessive) in the corresponding loci of a pair of homologous chromosomes. The condition is **homozygous**.

**hood:** See **laminar air flow cabinet**.

**hormone (plant hormone or phytohormone):** A substance naturally produced by plants which, in small amounts, regulates physiological activity, growth and development of cells, tissues and organs. This definition includes the naturally occurring auxins, gibberellins, abscisic acid, cytokinins, ethylene and hypothetical substances such as vernalin and florigen. Laboratory-synthesized substances identical to any of the naturally occurring ones are also **hormones**. Synthetically produced substances, chemically similar but not identical to natural hormones, which have hormone-like effects are synthetic hormone analogs. These include substances like benzylaminopurine (*N*-(phenylmethyl)-1*H*-purin-6-amine), kinetin (*N*-(2-furanylmethyl)-1*H*-purin-6-amine) and the many compounds which have auxin-like effects. Not included are the wide variety of herbicides ((2,4-dichlorophenoxy)acetic acid is an example), the effects of which are mostly traumatic. Hormones may be included singly or in combination and at various concentrations in plant tissue culture media to induce the desired effect on tissue cultures.

**horticulture:** An agricultural branch concerned with garden, orchard and ornamental plants grown for food, medicine or aesthetics.

**host specific toxin:** A metabolite produced by a pathogen which has a host specificity equivalent to that of the pathogen. Such toxins are utilized in in vitro selection experiments to screen for tolerance or resistance to the pathogen.

**hybrid:** **1.** The progeny ($F_1$, heterozygote) of a cross between two individuals different in one or more heritable characters (genes). The process of crossing the two individuals is **hybridization**. **2.** In genetic studies of relatedness of organisms, hybridization suggests reannealing of single stranded DNA and/or RNA prepared from two different organisms, forming complementary DNA molecules.

**hybrid selection:** The process of choosing plants possessing desired characteristics from among a hybrid population; identification and isolation of those desirable for further studies, propagation, etc.

**hybrid sterility:** The inability of some hybrids to produce viable gametes due to the absence of homologous chromosome pairs. If polyploidy can be induced, this condition may be overcome.

**hydathode:** A passive or active secretory structure through which water and other substances are exuded from the leaf or modified leaf surface. Passive secretion occurs through water pores, which are modified stomata, in the process of guttation and active secretion occurs through glandular organs (trichomes). Hydathodes are usually situated on vein endings and are linked to the tracheary elements.

**hydrate:** To add or incorporate water. The process is **hydration**.

**hydrochloric acid or muriatic acid (HCl mw 36.46):** A solution of hydrogen chloride in water. A strong acid used to lower the pH of media and to dissolve cytokinins and gibberellins in plant tissue culture work.

**hydrogen (H aw 1.00797 an 1):** A light, inflammable, colorless elemental gas. With oxygen it forms water and occurs in all living things. An essential major nutrient, it is obtained by plants from water in the process of photosynthesis.

**hydrogen peroxide ($H_2O_2$ mw 34.02):** Used as a disinfecting agent for plant material and as an oxidizing or bleaching agent. It is sold in dilute form (3% solution) suitable for disinfecting purposes.

**hydrophyte:** A plant adapted to water or very wet habitats.

**hydroponics:** The technique of cultivating plants (sometimes supported on an inert substrate) in circulated, aerated nutrient solutions containing essential nutrient elements. It is also known as soilless culture or water culture. Hydroponics is used in the commercial production of vegetables and flowers.

**[S-(Z,E)] - 5 - (1-hydroxy-2,6,6-trimethyl-4-oxo-2-cyclohexen-1-yl) - 3 - methyl-2,4-pentadienoic acid:** See **abscisic acid**.

**hygroscopic:** The ability of specific compounds to absorb water.

**hyperplasia:** An abnormal increase in cell number causing irregular swelling or growth. Hyperplasia is usually a disease or other stress-induced response.

**hyperploid:** A cell or organism with one or more extra chromosomes or chromosome segments in their complements; as hyperhaploid or hyperdiploid. The antomym is hypoploid. See **mitotic nondisjunction**.

**hypoploid:** A cell or organism with one or more deficient chromosomes or chromosome segments; as hypohaploid or hypodiploid. The antonym is hyperploid. See **mitotic nondisjunction**.

**hypertonic:** A solution whose osmotic potential is less than that of living cells; causing water loss, shrinkage or plasmolysis of cells.

**hypertrophy:** An abnormal increase in cell size causing irregular swelling or growth. Hypertrophy is usually a disease or other stress-induced response.

**hypha, pl. hyphae:** One of many threads of fungus cells that make up the fungal mat or mycelium.

**hypochlorite:** A salt of hypochlorous acid (**sodium hypochlorite, potassium hypochlorite** or **calcium hypochlorite**). All are oxidizing agents used for disinfecting and for bleaching.

**hypochlorous acid (HClO):** An acid existing only in solution whose salts (**hypochlorites**) have uses in disinfecting and bleaching.

**hypocotyl:** The part of the embryo or seedling stem (the embryonic stem) between the cotyledons and the radicle.

**hypoplastic:** Refers to an abnormal reduction in plant growth or development (dwarfing, stunting) resulting from disease or nutritional stress.

**hypotonic:** A solution whose osmotic potential is greater than that of living cells; causing swelling and turgidity of cells.

# I

**I:** The chemical symbol for the element **iodine**.

**IAA:** See **1*H*-indole-3-acetic acid**.

**IBA:** See **1*H*-indole-3-butanoic acid**.

**illuminate:** To supply or brighten with light. Illumination is an absolute requirement for most kinds of plant tissue cultures. Fluorescent lights are most commonly employed, the intensity of which is dependent on the light source and the requirements of the culture. The process is **illumination**, which is also the total visible radiation on a surface.

**imbibe:** To absorb liquid into a solid causing swelling; as the colloids agar or gelatin and seeds absorb water. The process is **imbibition**.

**Imperial units:** The British unit system based on the yard, pound and gallon. It is currently being replaced by SI units for scientific use and metric units for general use.

**in situ:** In the natural, original place or position; as in the location of the explant on the mother plant prior to excision.

**in vitro:** (Latin: "in glass"). Experimentation on organisms or portions thereof in glassware or culture; growing under artificial conditions as in tissue culture. The antonym of in vivo.

**in vitro pollination:** Pollination performed aseptically in vitro. Pollen is applied directly to ovules in an attempt to overcome various types of prezygotic incompatibility which otherwise inhibit fertilization.

**in vivo:** (Latin: "in life"). Experimentation on organisms under natural conditions within intact living organisms. The antonym of in vitro.

**incandescent:** Refers to light or light bulbs emitting light (white or red) by electrically heating an element (filament).

**incompatibility:** **1.** Selectively-restricted mating competence which limits fertilization; as lack of proper functions by otherwise normal pollen grains on certain pistils, a condition that may be caused by a variety of factors. **2.** A physiological interaction resulting in graft rejection or failure.

**incubate:** To maintain under conditions favorable for development, often in an **incubator**. The process is **incubation**.

**incubator:** An apparatus providing controlled environmental conditions, (light, photoperiod, temperature, humidity, etc.) suitable for (**incubating**) plants or plant cultures. This term is sometimes used as a synonym for cabinet (growth cabinet) but usually implies a greater degree of environmental control.

**indeterminate growth:** Refers to unlimited growth potential; as some apical meristems can produce unrestricted numbers of lateral organs for an indefinite period.

**indexing:** Any of several procedures for demonstrating the presence of specific viruses in suspect plants. It depends upon serological assay or the transfer of some part of this plant (such as buds, scions, sap, etc.) to an **indicator plant**.

**indicator plant:** A plant showing a reaction (symptoms) to an abiotic or biotic (pathogenic) factor used to assist the detection and identification of an environmental factor or pathogen.

**indirect embryogenesis:** Embryoid formation on callus tissues derived from zygotic or somatic embryos, seedling plant or other tissues in culture. The antonym is direct embryogenesis.

**indirect organogenesis:** Organ formation on callus tissues derived from explants. The antonym is direct organogenesis.

**1*H*-indole-3-acetic acid (IAA, $C_{10}H_9NO_2$ mw 175.18):** A naturally occurring plant growth hormone and the principal plant auxin. It is commonly used in plant tissue culture media. It dissolves in base (KOH or NaOH ca. 1 M). It is unstable to light so is usually stored in the dark.

**1*H*-indole-3-butanoic acid (IBA, $C_{12}H_{13}NO_2$ mw 203.23):** A naturally occurring plant growth hormone of the auxin type. It is synthetically produced and commonly used in plant tissue culture media, and in horticulture to promote rooting of cuttings. It dissolves in base (ca. 1 M KOH or NaOH).

**induce: 1.** To initiate a response (developmental, physiological, etc.); as an organ or structure. **2.** To cause variation or mutation. The process is **induction**.

**induction media: 1.** Media that can induce organs or other structures to form. **2.** Media that will cause variation or mutation in the tissues exposed to it.

**inert:** Physiologically neutral or immobile; as inert support structures contribute nothing chemically to plants but only function to support plant parts.

**infection: 1.** The entry and establishment of a pathogenic or parasitic organism (agent) into a host plant causing a disease or other harmful condition. Once the host shows symptoms it is said to be **infected** and may go through an **infectious** phase, during which it can spread disease to other organisms. **2.** Contamination with impurities or harmful properties; pollution.

**infestation:** Occupation by threatening numbers of insects, mites or potential disease agents; as mite **infestations** in incubators. An **infesting** organism can cause contamination of tissue cultures. This should not be confused with infected, as infection is not necessarily implied.

**infiltrate:** To introduce or permeate liquid into plant pores or spaces (tissue); as through applying a vacuum, then releasing it, during the disinfecting procedure. The process is **infiltration**.

**inflammable:** Easily ignited; as is ethanol. Synonym is **flammable**.

**inflorescence:** A complete flower cluster, including the axis and bracts; the arrangement of flowers on the floral axis.

**infrared gas analyzer (IRGA):** An instrument for measuring the proportion of a particular gas in a mixture; as changes in $CO_2$ concentration can be monitored to evaluate a plant's photosynthetic or respiratory activity.

**inhibit:** To prohibit, restrain or check free activity or function of a process or reaction; as toxic (**inhibitory**) substances in the medium may prevent culture growth. The process is **inhibition**.

**initial:** **1.** A cell with the capacity to start (**initiate**) or perpetuate growth. Such meristematic cells (**initials**) divide, perpetuating themselves and forming new body cells. Initials are present in the apical meristem and vascular cambium and from them specialized tissues develop. **2.** Refers to conditions existing at the beginning of a process, such as the concentration of medium components when a culture is begun.

**initiation:** **1.** Early steps in culture growth. **2.** Early formative steps in organogenesis in culture. **3.** Early stages of biosynthesis.

**inoculate:** **1.** To deliberately introduce something into; as **inoculum** is placed into (or onto) medium to initiate a culture. The process is **inoculation**, but should not be confused with contaminate. **2.** Vaccinate.

**inoculum:** **1.** Material introduced (**inoculated**) onto or into a host or medium. **2.** Potential inoculum implies that it is infective and may by chance result in natural **inoculation** of a host.

**inoculum size:** A critical minimum volume (minimum inoculum size) is necessary to initiate some culture growth due to the diffusive loss of cell materials into the medium. Inoculum size depends on medium volume and culture vessel size, and affects the subsequent culture growth cycle. Use of conditioned medium, that is medium that has supported culture growth, can decrease the minimum inoculum size.

**inorganic:** Not organic, having no reduced carbon or carbon compounds. Antonym of organic.

**inorganic salts:** Salts of mineral origin possessing no reduced carbon.

**inositol; myo, meso, or i-inositol or hexahydroxycyclohexane ($C_6H_{12}O_6$ mw 180.16):** A widely distributed stereoisomeric sugar alcohol in plants and essential in animals. Inositol is included by many in the B complex vitamin group. It is involved in the synthesis of phospholipids, cell wall pectins and membrane systems in cell cytoplasm. It is commonly added to nutrient medium used for plant tissue culture (ca. 100 mg/l) for its growth promoting effects.

**insecticide:** A chemical substance used to kill insects or control their populations.

**insertion:** **1.** The process of **inserting**. **2.** The position of anatomical structure attachment; as the parts of a flower.

**insertion element:** Generic term for DNA insertion sequences found in bacteria capable of genome insertion. Postulated to be responsible for site-specific phage and plasmid integration.

**instability:** A lack of steadiness, random type variation (developmental noise); as some cell lines loose certain characteristics or functions in culture due to genetic instability.

**integument:** Surrounds the ovule of seed plants; one in the gymnosperms, two in the angiosperms. The **integuments** form the testa (seed coat) after fertilization.

**intercalary meristem:** An internodal meristem, situated between differentiated tissues, that produces cells perpendicular to the growth axis causing internode elongation.

**intercellular:** Between cells.

**intergeneric hybrid:** A hybrid resulting from crossing species of two or more genera.

**internode:** The portion of the stem between one node, the point of leaf or branch attachment, and the next.

**interphase:** The interval between one mitotic or meiotic cell division and the next. Protein synthesis, chromosome duplication and other somatic activities occur during interphase.

**interspecific:** Between populations; as cybrids may result from fusion between cells of different populations.

**intracellular:** Within or through cells.

**intrageneric:** A type of hybrid resulting from a cross between species within one genus.

**intraspecific:** Within species; as cell fusion between members of the same population can give rise to cybrids or hybrids.

**iodine (I aw 126.9045 an 53):** A volatile blackish crystalline solid, nonmetallic, halogen element. This microelement is usually added to plant tissue culture media as **potassium iodide (KI)**. It is sometimes used as an antiseptic or germicide. It is also used as a starch indicator (Gram's Iodine) in solution with potassium iodide; prepared by adding 300 ml distilled water to 1.0 g I and 2.0 g KI.

**ion:** An electrically charged particle or group of atoms which has gained or lost one or more electrons; as salts are composed of atoms which dissociate when they are dissolved in water into negatively (**anions**) or positively (**cations**) charged particles.

**ion exchange:** A water purification procedure in which ions are removed from water by passing it through a resin bed. Also a powerful analytical tool for purifying biological molecules and macromolecules.

**ionizing radiation:** High energy, protoplasm-injuring radiation that can break covalent chemical bonds or remove electrons from atoms, attaching them to other atoms producing charged ion pairs. This radiation includes u.v., x-rays, $\gamma$-rays and $\beta$-particles.

**2iP or IPA:** See ***N*-(3-methyl-2-butenyl)-1*H*-purin-6-amine.**

**IRGA:** See **infrared gas analyzer**.

**iron (Fe aw 55.85 an 26):** A metallic element required in chlorophyll synthesis and a constituent of several cellular enzymes involved in photosynthesis and respiration. A deficiency of iron leads to chlorosis. Iron salts are routinely added to nutrient solutions for plant tissue culture; as ferric chloride, ferric citrate, ferric phosphate, ferric sulphate, ferric tartrate, ferrous sulphate or ferrous tartrate. It is sometimes necessary to chelate iron salts with an agent such as $Na_2EDTA$.

**irradiate:** **1.** To illuminate. **2.** To emit or expose to waves of light, heat or nuclear emissions. **3.** To treat by exposure to radiation. The process is **irradiation**.

**irradiance:** The total radiation on an exposed surface.

**isodiametric:** Having equal diameters. Used to describe plant cells.

**isoenzyme:** See **isozyme**.

**isogenic line:** Cells or tissues possessing an identical genotype.

**isograft or syngraft:** A graft or transplant among isogenic (genetically identical) individuals; as on the same organism.

**isolate:** **1.** To set apart, place or keep separate from others and make a pure culture of; as to place an explant into culture. **2.** To maintain healthy cultures or plants away from potential inoculum; or maintain infected material (contaminated cultures) away from potential hosts or clean cultures. The process is **isolation**. **3.** A unique source may be referred to as an isolate.

**isolation medium:** A medium suitable for explant survival and development. It may be synonymous with Stage I medium, or contain antioxidant(s), reduced hormone concentration, bacterial indicators or other addenda, and preceed Stage I culture.

**isopentenyladenosine:** See ***N*-(3-methyl-2-butenyl)-1*H*-purin-6-amine**.

**isopropanol ($C_3H_8O$ mw 60.09):** An alcohol that is sometimes used for disinfecting purposes as a less costly alternative to ethanol.

**isotonic or isoosmotic:** With the same osmotic potential, the same molar concentration of a solution; as protoplasts will not survive if the medium they are suspended in is not isotonic.

**isozyme or isoenzyme:** An enzyme (protein) with the same function and sometimes the same activity, but with different structure, to another enzyme present in a cell or organism. Isozyme analysis involves electrophoresis, and may confirm the production of somatic hybrids, distinguish between related species or cultivars and has many other potential uses.

# J

**Jiffy pots:** A brand name for peat pots that are sometimes used for ex vitro transplantation.

**j/m² or joule/square metre:** A commonly used unit of light measurement.

**juvenile:** The immature, usually non-reproductive (reproductively incompetent), sometimes phenotypically distinct phase of plant growth. **Juvenility** should not be confused with young in age although young plants may also be in a juvenile phase. Juvenile phase tissue has repeatedly been found to perform better in culture than adult phase tissue. It is more likely to grow and to form callus, embryoids, shoots, etc.

# K

**K:** The chemical symbol for the element **potassium**.

**Kaput:** A brand name for a plastic cap used as a test tube closure.

**karyogamy:** Nuclear fusion. The essential feature of sexual reproduction.

**karyokinesis:** Nuclear division, as distinct from and usually preceeding cytoplasmic division (cytokinesis).

**karyotype:** Chromosome size, shape, number and other characteristics of the typical chromosome set of an individual species. The karyotype may be depicted diagramatically at mitotic metaphase in a **karyogram** (idiogram).

**Kimkap:** A brand name for a plastic cap used as a test tube closure.

**kinetin (K, sometimes Kn):** See ***N*-(2-furanylmethyl)-1*H*-purin-6-amine**.

**kinin:** The original class name for substances promoting cell-division to which the prefix "cyto" has been added (**cytokinins**) to distinguish them from **kinins** in animal systems.

**Knop, J. (1817–1891):** A macroelement solution based on his soil analysis is frequently employed in soilless culture systems and as a fertilizer. It is most commonly used at double the original (Knop, 1865) concentrations.

**Kogl, F., A.J. Haagen-Smit and H. Erxleben (1934):** Identified the first plant hormone, 1*H*-indole-3-acetic acid (IAA).

**Krebs cycle:** See **tricarboxylic acid cycle**.

# L

**lactalbumin hydrolysate:** A common undefined organic addendum to plant tissue culture media.

**lactoflavin:** See **riboflavin**.

**lactose ($C_{12}H_{22}O_{11}$ mw 342.31):** A disaccharide of glucose and galactose present in milk.

**lag phase:** **1.** A latent period; the state of apparent inactivity preceeding a response. **2.** Used to describe the first of five growth phases of most batch propagated cell suspension cultures in which inoculated cells in fresh medium adapt to the new environment and prepare to divide.

**lamina:** A leaf blade.

**laminar air flow cabinet or hood:** A structure in which a uniform flow of filtered air prevents settling of particles in the work area. The air is filtered through a prefilter (furnace-filter quality), then a high efficiency particulate air (HEPA) filter that strains out particles greater than 0.3 $\mu$m. These hoods are commonly employed for aseptic plant tissue culture manipulations and are designed to accommodate one to several persons, depending on the model.

**lateral:** Buds, shoots or other structures emerging from one side (axillary) of an organ; as opposed to terminal.

**lateral bud or axillary bud:** A bud located in the axil of a leaf.

**lateral meristem:** A meristem giving rise to secondary plant tissues; as the vascular and cork cambia. Less often the term refers to axillary meristems.

**layering:** **1.** Covering stems, runners or stolons with soil causing adventitious roots to form at the nodes enabling propagation by rooted cuttings. This procedure is used commercially to propagate many plants, such as the brambles. **2.** In vitro layering involves the horizontal placement on agar of cultured shoots (with or without leaves) or nodal segments to promote axillary bud proliferation.

**leaf:** A green organ arising laterally or from the shoot apex of a plant. They are usually dorsiventral in structure, simple or compound. A leaf blade is supported by a petiole, or when petioles are absent, **leaves** are sessile. When the petiole is blade-like, and the blade is absent, they are called phyllodes. Leaves are the photosynthesizing, respiring and transpiring portion of most plants. They are common explant sources for plant tissue culture, especially as buds. Leaves developed in vitro tend to be phenotypically different from those developed under greenhouse or field conditions. This is due to aspects of the culture environment including relatively low light intensity and relatively high temperature and humidity.

**leaf buttress:** Located below the meristematic dome (meristem), this structure appears as a small lateral protrusion which develops into a leaf primordium.

**leaf primordium:** The initial stage of leaf formation consisting of a lateral outgrowth below the meristematic dome (meristem).

**lecithin:** A naturally occurring, choline-containing phospholipid present in animal and plant tissue. Chemically **lecithins** are similar to fats, but they also contain phosphorus and nitrogen.

**levulose or laevulose:** See **fructose**.

**leucoplast:** A colorless plastid, as are elaioplasts, aleuroplasts and amyloplasts.

**light:** Electromagnetic radiation with wavelengths in the visible range, 400–700 nm. Light energy fuels photosynthesis by green plants. Light quality, intensity and photoperiod are usually controlled in incubators and cabinets used for plant tissue culture and have been empirically determined. Light is a requirement of many but not all cultures. Usually cool white fluorescent tubes are used, sometimes supplemented with incandescent or warm white fluorescent. The usual light intensity employed ranges from 0 to ca. 125 $\mu molm^{-2}s^{-1}$ (0 to 10,000 lux). The usual photoperiod is 12 to 16 hours, although some cultures are incubated in total (24 hour) light.

**lignin:** A complex phenolic compound, considered to be a polymer of phenyl propane units, deposited in cell walls of xylem vessels, tracheids and sclerenchyma and lending strength and rigidity. It is the major constituent of wood, forming 25–30% of the wood of trees. The process of deposition of lignin in cells walls is **lignification**.

**limiting requirement or factor:** An environmental variable whose absolute level at a given time limits the growth or other activity of an organism.

**lineage:** The line of common descent; as the cell line originating from a single cell plated in vitro.

**linear phase:** The constant increase in cell number following the exponential growth phase and preceeding the deceleration phase in most batch suspension cultures.

**$\beta$(1–4) linkage:** An oxygen bridge linking the number 1 carbon atom to the number 4 carbon of another molecule.

**liposome:** A spontaneously formed, layered lipid vesicle in aqueous medium used as a DNA vector in cell hybridization work.

**liquid culture:** The culture of plant cells on liquid medium, in suspension or on supports. Cultures are either held steady (stationary culture) or are shaken (agitated culture or shake culture).

**liquid medium:** Medium not solidified with a gelling agent. **Liquid media** are used for suspension cultures and for a wide range of research purposes. They are also useful for Stage I cultures in some micropropagation protocols. Usually a support structure or wick is used to hold the tissue above the nutrient medium.

**liquid nitrogen:** Nitrogen gas which has been condensed to a liquid and has a boiling point of −195.79°C. Immersion is utilized as a means of rapidly bringing tissues to ultra low temperature. Used in cryopreservation and some anatomical studies.

**litre (l):** An S.I. metric volume measure equal to one $dm^3$.

**lithotroph:** See **autotroph**.

**litmus:** A pH indicator paper (range 4.5–8.3) made from an extracted lichen pigment. It turns red in acidic and blue in alkaline solutions, but is not a precise method of pH measurement.

**Lloyd, G. and B. McCown (1980):** See **woody plant medium**.

**locule:** A cavity or chamber; as within a pollen sac or ovary.

**lux:** A unit of light measurement (0.0929 foot candles) once widely employed in horticultural and physiological studies but now largely supplanted by photosynthetically active radiation (PAR) units, including $\mu molm^{-2}s^{-1}$ ($\mu molEm^{-2}s^{-1}$) and $Wm^{-2}$.

**lyophilize or freeze dry:** To freeze rapidly then dehydrate under high vacuum. The process is **lyophilization**.

**lysis:** Cell rupture or destruction, as through enzymatic action.

# M

**$M_1$:** Designates a **mutant** generation; as following a mutational event such as colchicine application. **$M_2$**, **$M_3$** are the second and third generations from crosses between the **$M_1$** and **$M_2$** respectively.

**macerate:** To disintegrate or separate tissues through cutting, soaking, enzymatic or other action, resulting in cell dissociation.

**Macerase:** A brand name for pectinase. Pectinase is useful in the isolation of intact cells and protoplasts of higher plants.

**macroelement or macronutrient element or major element:** One of the main plant nutrient elements essential for plant growth, and required in greater amounts than the others (microelements or micronutrient elements). The **macroelements** are carbon (C), oxygen (O), hydrogen (H), nitrogen (N), phosphorus (P), potassium (K), calcium (Ca), magnesium (Mg) and sulphur (S). Sometimes iron (Fe) is included, more often it is classed with the microelements. Six macroelements (N, P, K, Ca, Mg and S) are added to nutrient media employed for plant tissue culture in the form of salts and are usually required in concentrations above 0.5 mM/litre.

**magnesium (Mg aw 24.312 an 12):** A silvery white, metallic element. As a macroelement it is a component of chlorophyll and many enzyme systems. Its deficiency leads to chlorosis, necrosis, stunting or other symptoms. It is a constituent of nutrient media for plant tissue culture, usually added as **magnesium chloride**, **magnesium nitrate** or **magnesium sulphate**.

**magnesium porphyrin:** A substituted tetrapyrrole in which the four nitrogen atoms are coordinated to a magnesium atom.

**magnetic stirrer:** An apparatus used for stirring solutions. A magnetic stir bar is placed at the bottom of the container and is twirled in the solution that requires mixing by an electrically driven, rotating magnet below the platform of the apparatus.

**maintain:** To sustain, provide for or support; as cultures are kept **maintained** on fresh medium in incubators. The process is **maintenance**.

**major element:** See **macroelement**.

**malachite green ($C_{23}H_{25}N_2Cl$ mw 364.66):** A chemical used as a seed disinfectant in agriculture. It is included in some tissue culture media as an antiviral agent but has received mixed reviews.

**male sterile or cytoplasmic male sterile:** Plants incapable of viable pollen production.

**malt extract:** An extract from dried and ground, germinated grain seed (usually barley, sometimes rice or corn). It is an undefined constituent of some plant tissue culture media.

**maltose or malt sugar ($C_{12}H_{22}O_{11}$ mw 342.31):** A disaccharide of glucose units widely distributed in plants and used as a carbohydrate source in some plant tissue culture media.

**manganese (Mn aw 54.94 an 25):** A hard, reddish, metal element. As an essential microelement it is a component of chloroplast membranes and some enzyme systems. It is a common constituent of nutrient media for plant tissue culture and is added as **manganous chloride** or **manganous sulphate**.

**mannitol ($C_6H_{14}O_6$ mw 182.17):** A sugar alcohol widely distributed in plants and often employed as a nutrient and osmoticum in plant tissue culture work; as in suspension medium for plant protoplasts.

**mannose ($C_6H_{12}O_6$ mw 180.16):** A hexose component of many polysaccharides and mannitol. It is occasionally employed as a carbohydrate source in plant tissue culture media.

**marginal meristem:** A meristem along the leaf primordial margin that forms the mesophyll and epidermal tissues of the leaf blade.

**mass meristem:** A meristematic tissue which increases in volume through cell divisions in various planes.

**maternal inheritance:** Inheritance controlled by extrachromosomal (cytoplasmic) hereditary determinants.

**matric potential:** A water potential component, always of negative value, resulting from capillary, imbibitional and adsorptive forces.

**mature:** Fully differentiated and functionally competent cells, tissues or organisms.

**MDA:** See **microdroplet array technique**.

**medium:** The substrate for plant growth; as nutrient solution, soil, sand, etc. This is a general term for the liquid or solidified formulation upon which plant cells, tissues or organs develop in plant tissue culture.

**medium formulation:** One of many available, usually empirically derived formulas that are used in plant tissue culture. The formulas commonly contain the macroelements and microelements (high and low salt formulas are available), some vitamins (B vitamins, inositol), hormones (auxin, cytokinin and sometimes gibberellin), a carbohydrate source (usually sucrose or glucose) and often other substances such as amino acids (the most common addition is glycine) or complex growth factors. Media may be liquid or solidified with agar, the pH is adjusted (ca. 5–6) and the solution is sterilized (usually by filtration or autoclaving). Some formulations are very specific in the kind of explant or plant species that can be maintained, some are very general.

**megaspore or macrospore:** A haploid (n) spore (usually the larger of two spore types) developing into a female gametophyte in heterosporous plants.

**meiosis:** Reduction division of a zygote; two orderly divisions during which chromosomes are reduced from diploid (2n) to haploid (n) and genetic recombination occurs. Meiosis results in gene segregation and the production of gametes or spores in sexual organisms.

**meiotic analysis:** Used to analyze chromosome-pairing relationships.

**Melchers, G.L. and L. Bergmann (1959):** Were the first to culture haploid tissue other than pollen.

**meniscus:** The curved surface at the top of a liquid column. The bottom of the meniscus is used as a guide when liquid volumes are measured in calibrated containers.

**mercuric chloride or mercury bichloride ($HgCl_2$ mw 271.52):** This compound was once commonly used as an antiseptic and fixing agent (0.05-0.10%) for plant material. However, it is highly toxic and is now little used for this purpose.

**mericlinal:** Refers to a chimera with tissue of one genotype partly surrounded by that of another genotype.

**mericlone:** A meristem tip source clone. This term can be synonymous with calliclone or may not involve callus.

**meristem:** A localized region of continuing mitotic cell division (**meristematic** cells), of protoplasmic synthesis and tissue initiation. From these undifferentiated tissues new cells arise that differentiate into specialized tissues. **Meristems** are located at the apical (shoot and root tips), axillary, marginal or lateral (cambia) and other growing points. In addition, in vitro meristems may occur within more or less differentiated callus tissue and are termed **meristemoids**. The **meristematic** dome without any leaf primordial tissue is sometimes used as an explant in virus elimination work but usually a meristem tip (meristematic dome plus one pair of leaf primordia) is the explant.

**meristem culture:** The culture of meristems; meristematic dome tissue without adjacent leaf primordia or stem tissue. In practice this usually means meristem tip culture. This may more generally imply the culture of meristemoidal regions of plants or of meristematic growth (associated with or sharing the characteristics of the meristem) in culture.

**meristem tip:** A common explant comprised of the meristem (meristematic dome) and usually one pair of leaf primordia. Also refers to apical (apical meristem tip) or lateral (lateral or axillary meristem tip) origin. The term meristem tip is often confused with the term shoot tip, which is much larger and usually has more immature leaves and some stem tissue.

**meristem tip culture:** Cultures derived from meristem tip explants. They are excised for virus elimination or axillary shoot proliferation purposes, less commonly for callus production.

**meristemming:** The utilization of or process of utilizing meristem tips for explants. This term is often used to refer to micropropagation via axillary shoot proliferation.

**meristemoid:** A cluster of small, isodiametric **meristematic** cells within a **meristem**, or cultured tissue, with the potential for developmental (totipotential) growth. These cells have a dense microvacuolated cytoplasm and a high nucleo-cytoplasmic ratio. They may give rise to plant organs (shoots, roots) or entire plantlets in culture.

**mesocotyl:** The internode between the coleoptile and the scutellum in grass embryos and seedlings.

**meso-inositol:** See **inositol**.

**mesophyll:** Thin-walled, chloroplast-containing (chlorenchyma), photosynthesizing parenchyma tissue composing the bulk of the leaf between the upper and lower epidermis. The palisade and spongy parenchyma are the two mesophyll cell types of dicots. The mesophyll may also contain collenchyma and sclerenchyma cells, which support the veins.

**mesophyte:** A plant with moderate moisture requirements; as compared with a hydrophyte or xerophyte.

**messenger RNA:** See **ribonucleic acid, messenger**.

**metabolism:** The chemical or energy processes involved in obligatory life activities occurring in all living cells, organs and organisms. Broadly speaking this involves catabolism, the breakdown of more complex to simple organic compounds with liberation of available energy for cell or organ activities, and anabolism, the synthesis of simple to complex organic compounds, using available energy from catabolism or, among autotrophs, from non organic external sources.

**metabolite:** A substance which participates in the process of metabolism, most of which are synthesized by the organism, others taken in from the environment. Autotrophic organisms take in inorganic **metabolites**, as water, $CO_2$, nitrates and some trace elements. Heterotrophic organisms take in these inorganic metabolites as well as a wide range of organic metabolites including some vitamins and amino acids.

**metaphase:** The mitotic phase during which the chromosomes (pairs of chromatids) align on the equatorial plane.

**metaphloem:** See **phloem**.

**metaxylem:** See **xylem**.

**methionine (Met, $C_5H_{11}NO_2S$ mw 149.21):** An amino acid precursor in ethylene synthesis and occasionally added to plant tissue culture media.

**methyl methanesulfonate (MMS, $C_2H_6SO_3$ mw 110.13):** A frequently used, very potent chemical mutagen which acts by adding methyl groups to guanine and subsequently causes base pairing errors as it binds to adenine.

**1-methyl-3′-nitro-1-nitrosoguanidine (MNNG or NTG, $C_2H_5N_5O_3$ mw 147.094):** A very potent chemical mutagen.

**2-methyl-4-(1*H*-purin-6-ylamino)-2-buten-1-ol:** See **zeatin**.

***N*-(3-methyl-2-butenyl)-1*H*-purin-6-amine or isopentenyladenosine (2iP or IPA, $C_{10}H_{13}N_5$ mw 203.20):** A synthetic cytokinin similar in structure to zeatin and commonly used in plant tissue culture media. It dissolves in acid (HCl ca. 1 M).

**Mg:** The chemical symbol for the element **magnesium**.

**mho:** A unit of electrical conductivity. The conductance per $cm^3$ with a 1 volt potential allows the passage of 1 ampere current per $cm^2$. The reciprocal of the resistance unit ohm.

**microbe:** A microscopic (may be pathogenic) organism (microorganism).

**microclimate or microhabitat:** The climate in the immediate vicinity or surrounding an organism or its parts.

**microcutting:** A tiny cutting; as an explant such as a meristem tip removed for culture with the use of a dissecting microscope.

**microdroplet array or multiple drop array (MDA) or hanging droplet technique:** Introduced by Potrykus, Harms and Lorz (1979), this technique is used to evaluate large numbers of media modifications, employing small quantities of medium into which are placed small numbers of cells. Droplets of liquid culture (medium and suspended cells or protoplasts) are arranged on the lid of a petri dish, inverted over the bottom half of the dish containing a solution with a lower osmotic pressure, and the dish is sealed. The cells or protoplasts form a monolayer at the droplet meniscus and can easily be examined.

**microeinstein ($\mu$E):** A photometric measurement consisting of 0.000001 mole photons. The usual units are $\mu molm^{-2}s^{-1}$ or $\mu Em^{-2}s^{-1}$.

**microelement or micronutrient element or minor element:** One of the nutrient elements necessary in small (trace) amounts for plant growth. The usual concentration required is below 0.5 mM/l. These include boron (B), molybdenum (Mo), manganese (Mn), copper (Cu), zinc (Zn) and usually iron (Fe). Several additional elements are required by some plant species but not others, such as sodium (Na), iodine (I), cobalt (Co), chlorine (Cl) and vanadium (Va). The term micronutrient may also include plant nutrients required in trace amounts, such as vitamins.

**micrograft:** An in vitro graft. This procedure involves placing a meristem tip or shoot tip explant onto a decapitated rootstock, that has been grown aseptically from seed or micropropagated. The purpose may vary, from virus elimination, using meristem tips and virus-free rootstocks, to viral assays or to expedite grafting that is usually done in the greenhouse. The explanted scion is placed directly onto the cambium of the severed stock or inserted into a T or inverted T-shaped epidermal incision on the stock. This technique has been used to eliminate viruses, viroids and mycoplasmas from *Citrus* species.

**microlitre ($\mu$l):** An SI unit of measurement equal in volume to one thousandth (0.001) of a millilitre ($10^{-3}$ ml).

**micrometer:** An instrument or grid used to measure small distances or angles; as a calibrated ocular scale (graticule) fits in the eyepiece (ocular) of a microscope and is used after calibration with a stage micrometer to measure the size of objects or specimens in the field of view.

**micrometre ($\mu$m):** An SI unit of measurement equal in length to one thousandth (0.001) of a millimetre ($10^{-3}$ mm). The former term was micron ($\mu$).

**micromho:** A unit of electrical conductivity ($10^{-6}$ mho or $1/10^{-6}$ ohm).

**micromole ($\mu$M):** One millionth (0.000001) of a mole ($10^{-6}$ M). The unit of concentration used for microelements, vitamins, hormones and some other organic addenda used in plant tissue culture media.

**micron ($\mu$):** The length of 0.001 millimetre ($10^{-3}$ mm). This unit has been replaced by the unit micrometre ($\mu$m).

**micronutrient element:** See **microelement.**

**microorganism:** A microscopic, unicellular, plant, animal or bacterial organism.

**microprecipitin test:** See **precipitin test.**

**micropropagation:** Refers to propagation in culture by axillary or adventitious means. It is a general term for vegetative (asexual) in vitro propagation. It sometimes refers specifically to axillary bud proliferation.

**micropyle:** The canal formed by extension of the integuments of the ovule and through which the pollen tube grows into the embryo sac. In the mature seed it appears as a minute pore in the seed coat through which water is imbibed when the seed begins to germinate.

**microspore:** A haploid (n) spore or immature pollen grain which develops into a male gametophyte in heterosporous plants. In some species these can develop in culture into haploid callus which can be induced to form haploid plants, or into embryoids in vitro without an intervening callus phase.

**microsporogenesis:** Refers to developmental stages of **microspore** formation in the anthers of angiosperms.

**microvacuole:** A minute vacuole which may be invisible to the light microscope but observable with the electron microscope.

**microwave oven:** An oven in which heating results from microwave radiation of the food or contents. Such ovens are commonly employed to melt agar in making up plant nutrient media and may be useful to sterilize liquids and some plastic or glass items.

**middle lamella:** The intercellular layer cementing adjacent primary cell walls together and made up of pectinaceous materials or lignin in woody tissues.

**midrib:** The major, central leaf vein.

**Miller, C.O. (1955):** The first to isolate kinetin (6-furfurylaminopurine), the first cytokinin, a breakdown product of DNA. This very important discovery permitted the replacement of coconut milk with kinetin enabling the culture of plant tissues on chemically defined media.

**milligram (mg):** An SI unit of measurement equal in weight to one thousandth (0.001) of a gram ($10^{-3}$)g.

**millilitre (ml) or cubic centimetre (cc):** An SI unit of measurement equal in volume to one thousandth (0.001) of a litre ($10^{-3}$l).

**millimetre (mm):** An SI unit of measurement equal in length to one tenth (0.1) centimetre or one thousandth (0.001) of a metre ($10^{-3}$ m).

**millimho:** One thousandth (0.001) of a mho ($10^{-3}$ mho) or $1/10^{-3}$ ohm.

**millimole (mM):** One thousandth (0.001) of a mole ($10^{-3}$ M). A unit of concentration used for macronutrient elements and organic nutrients in tissue culture media.

**Millipore filter:** The brand name for a type of membrane filter. These filters come in a range of pore sizes and are used for filtering gases or liquids.

**mineral:** Any of many homogeneous, usually solid inorganic substances present in the soil, occurring as the result of natural processes.

**minimum effective cell density:** The inoculum density below which the culture fails to give reproducible cell growth. The minimum density is a function of the tissue (species, explant, cell line) and the culture phase of the inoculum suspension. Minimum density decreases inversely to the aggregate size and division rate of the stock culture. It is important to know this ratio when plating suspension cultures for selection of single cell lines so that colonies will not overgrow one another prior to obtaining readily manipulable size.

**minimum inoculum size:** The smallest inoculum that can be successfully used for subculture.

**minor element:** See **microelement**.

**mite:** Animals classified in order Acarina, class Arachnida (with spiders). There are free-living and parasitic forms. **Mites** may infest plant tissue culture work areas and incubation facilities in search of sugars and so contaminate culture vessels and spread bacteria and fungi. **Miticidal** (acaricidal) shelf paper or sprays are useful in their control, as is rapid disposal of used or contaminated cultures.

**mitochondrion, pl. mitochondria:** Cytoplasmic organelle in eukaryotic cells, associated with intracellular respiration; containing Kreb cycle enzymes and those of oxidative phosphorylation. Mitochondria, which may reach 10,000 per cell, function in the production of ATP energy for cellular functions, so are sometimes termed the "powerhouse of the cell".

**mitosis:** The longitudinal doubling of chromosomes and separation of daughter chromosomes in the process of nuclear duplication and division (karyokinesis) that usually accompanies cell division (cytokinesis) into daughter cells in eukaryotic cells. The active **mitotic** stages are prophase, metaphase, anaphase, telophase and these are followed by interphase (the "resting" nucleus).

**mitotic analysis:** Examination of the mitotic division cycle. The purpose might be to count the number of chromosomes (at metaphase) or to calculate the rate of cell division, etc.

**mitotic index (MI):** The ratio of nuclei undergoing mitosis (including prophase) to total nuclei. MI = (Number of nuclei in mitosis/Total number of nuclei examined in the sample) $\times$ 100. A MI of 0.3 means that 30% of cells in the population are observed in mitosis.

**mitotic nondisjunction:** Occurs when sister chromatids fail to migrate to opposite poles of the cell during mitotic anaphase. The result is daughter cells with hyperploid and hypoploid chromosome counts.

**mixoploid:** Refers to cells with variable (euploid, aneuploid) chromosome numbers; as mosaic or chimaeral components that differ in chromosome number, or the result of a variety of mitotic irregularities.

**Mn:** The chemical symbol for the element **manganese**.

**Mo:** The chemical symbol for the element **molybdenum**.

**molar or molar solution:** One mole, or one gram molecular weight of a material (solute), dissolved in one litre of solution.

**mold:** Any fungus growth; as on plants or in cultures.

**mole (mol):** An SI unit of mass. One gram molecular weight of a substance; the number of particles is always Avogadro's number ($6.023 \times 10^{23}$) although the weight varies with the substance. The formula weight of a substance, in grams; the sum of the atomic weights of a compound in grams.

**molecular weight:** The sum of the atomic weights of the atoms in a molecule.

**molecule:** The smallest particle of matter that is chemically the same as the mass. The simplest molecules have only one atom; as does helium.

**$\mu molm^{-2}s^{-1}$:** Replaces the unit $\mu Em^{-2}s^{-1}$ for light measurement.

**molybdenum (Mo aw 95.94 an 42):** A hard metallic element which does not occur free in nature. As a microelement it participates in biochemical redox reactions. It is involved in nitrate reduction. It is added to plant tissue culture media most commonly as **sodium molybdate**.

**monocot or monocotyledon:** One of (the smaller) of the two angiosperm classes (**Monocotyledoneae**). Plants in this class are characterized by an embryo with a single cotyledon, parallel veined leaves, flower parts in threes or multiples thereof, a fibrous root system, no vascular cambium and stem vasculature of two or more rings or scattered closed bundles. There are more than 75,000 species in this class including grasses, orchids, lilies, etc.

**monoculture:** A one-crop agricultural system. An intensified monoculture could refer to a clonal (one genotype) agricultural system.

**mononucleotide:** A nucleic acid basic-building-block. Composed of a pentose, a phosphoric acid plus a purine or pyrimidine base.

**monoploid:** Possessing half the parental chromosome complement. For example, in tetraploid species **monoploids** are 2n.

**monopodial:** Indeterminate terminal bud growth, a growth habit shared by some orchids.

**monosaccharide:** A carbohydrate consisting of a single simple sugar unit such as glucose or fructose.

**monosomic:** A plant with one fewer than the normal chromosome complement. The condition is **monosomy**, a type of aneuploidy. A double monosomic is a polyploid with two chromosomes missing; if these are homologous chromosomes, the plant is nullisomic.

**Morel, G.M. (1960):** Showed that meristem tip culture could be used for the rapid clonal propagation of orchids.

**Morel, G.M. and G. Martin (1952):** Demonstrated that meristem tip culture was an effective means of recovering virus-free plants from infected plants.

**morphogenesis:** The anatomical (**morphological**) and physiological events involved in the growth and development (ontogeny) of an organism, in the formation of its characteristic organs and structures, or in regeneration.

**morphogenetic response:** The effect on the developmental history of a plant or its parts exposed to a given set of growth conditions or to a change in this environment.

**morphology:** The study and science of plant form, structure and development.

**morphosis:** Nonadaptive, usually unstable variation in **morphogenesis** associated with certain environmental changes. These phenocopies may mimic the effects of known mutations.

**mosaic:** **1.** An individual composed of cells differing in genotype or plasmotype. **2.** Leaves where discrete colored areas such as green, yellow or white form patterns which stand out on the surface. These symptoms indicate disease (virus) or a genetic mixture (chimera) which may be the result of mutations.

**mother plant:** A source plant or donor plant from which an explant used to initiate a culture is taken.

**mRNA:** See **ribonucleic acid, messenger**.

**MS:** An abbreviation for the **Murashige, T. and F. Skoog (1962)** medium formulation. MS is a defined medium formulated for tobacco callus culture and now used more than any other for the culture of a very wide range of plant species. It is characterized by higher levels of nitrogen (both nitrate and ammonium), potassium and calcium than most other plant culture media.

**Muir, W.H. (1953):** Demonstrated the growth promotion of single cells taken from suspension cultures on filter paper set above actively growing callus or "nurse" tissue (paper raft technique).

**Muir, W.H., A.C. Hildebrandt and A.J. Riker (1954):** Obtained the first suspension cultures by transferring callus fragments to agitated liquid medium.

**multinucleate:** A cell containing many nuclei.

**multiple drop array technique:** See **microdroplet array technique**.

**multiplication or proliferation:** **1.** Any tissue culture undergoing rapid growth. **2.** Commonly used to describe Stage II culture, in which consecutive subcultures promote an increase in the number of propagules. **3.** The media employed to induce culture proliferation (multiplication media).

**muriatic acid:** See **hydrochloric acid**.

**mutability:** The propensity of an individual's genes or genotype to undergo heritable mutation.

**mutagen:** A chemical or physical treatment or agent capable of producing genetic mutation. Causes insertion, deletion or alteration of a base or part of the nucleic acid chain (DNA). Artificial mutations can be induced in cells using chemicals or ionizing radiation. The process is **mutagenesis**. Electromagnetic and particle radiation, such as u.v., x-rays and $\beta$ particles all increase mutation frequency. So do chemical mutagens such as ethylmethanesulphonate, acridine, nitrous oxide and many others.

**mutant:** A genetically altered cell or plant variant showing one or more discrete heritable differences in physiology or morphology from the clone, or from the wild type and carrying mutated gene(s) (gene mutation) or chromosome(s) (chromosome mutation).

**mutation:** A sudden and permanent change in a hereditary character, or one marked by this change; as **mutational** or mutagenic events including exposure to mutagens like radiation, ultraviolet light or carcinogenic chemicals. The sequence of nucleotides within a gene are altered resulting in a change in the gene-specified polypeptide. Complex changes of this sort include duplications, deletions, inversions or translocations of portions of chromosomes. The production of mutants is mutagenesis. Mutation in the gametes, if not lethal, is inherited by the offspring. Somatic mutation (in body cells) is passed to all cells derived through mitosis.

**mutation breeding:** To (experimentally) introduce or remove a character from a cell or organism by exposure to mutagenic agents followed by screening for the desired attribute.

**mycelium:** The collective term for fungal hyphae. These are symptomatic of contamination in plant tissue cultures.

**mycoplasma:** A member of a genus of prokaryotic organisms (**Mycoplasma**) some of which cause plant diseases and are the smallest free-living microorganisms with ribosomes and both DNA and RNA. **Mycoplasmas** have a unit membrane but no rigid cell wall and can pass through bacteria-retaining filters. They can be eliminated from plants by thermotherapy alone or in conjunction with meristem or meristem tip culture.

**mycorrhiza:** Fungi which form a mutually beneficial association with the roots of a plant. The fungus enhances the ability of the roots to take up nutrients and water while the fungus obtains carbohydrates and vitamins from the plant. There are two main types of **mycorrhizal** associations. Ectotrophic or ectophytic mycorrhiza develop between a woody plant and a basidiomycete fungus. The hyphae invade the root tip, suppressing root hair formation and penetrate between cortical cells and around the surface of the roots, forming a mantle. In endotrophic or endophytic mycorrhizal associations the fungus lives in and between the root cortical cells and the association is usually obligate. These fungi are of various groups and form associations with both herbaceous and woody hosts.

**myo inositol:** See **inositol**.

# N

**n: 1.** The symbol indicating the haploid chromosome number. **2.** An abbreviation for number, indicating sample size.

**N: 1.** The chemical symbol for the element **nitrogen**. **2.** An abbreviation for the **Nitch, J.P. (1951)** medium formulation.

**2n:** The symbol indicating the diploid chromosome number.

**Na:** The chemical symbol for the element **sodium**.

**NAA:** See **1-naphthaleneacetic acid**.

**$NAA_m$ or NAD:** See **1-naphthaleneacetamide**.

**1-naphthaleneacetic acid or α-naphthaleneacetic acid (NAA, $C_{12}H_{10}O_2$ mw 186.20):** A synthetic hormone analog of the auxin type. It is frequently used in plant tissue culture media and in horticulture to promote rooting of cuttings. It dissolves in base (ca. 1M KOH or NaOH).

**α-naphthaleneacetic acid:** See **1-naphthaleneacetic acid**.

**1-naphthaleneacetamide (NAM or $NAA_m$ mw 185.14):** A synthetic hormone analog of the auxin type. It is sometimes used in plant tissue culture media. It dissolves in base (ca. 1M KOH or NaOH).

**(2-naphthalenyloxy)acetic acid or 2-naphthoxyacetic acid or β-naphthoxyacetic acid (NOA or βNOA, $C_{12}H_{10}O_3$ mw 202.20):** A synthetic hormone analog of the auxin type. It is sometimes used in plant tissue culture media and in horticulture to promote rooting of cuttings. It dissolves in base (ca. 1M KOH or NaOH).

**β-napthoxyacetic acid:** See **(2-naphthalenyloxy)acetic acid**.

**natural complex:** A complicated, non-synthetic, often undefined addendum in plant tissue culture media, such as orange or tomato juice or coconut milk.

**necrosis:** The localized death and discoloration of plant tissue. These **necrosed** areas are referred to as **necrotic lesions**.

**neoplasm:** Localized cell multiplication or tumor; a collection of cells which have undergone genetic transformation. These cells differ in structure and function from the original cell type.

**neoteny:** The occurrence (retention) of juvenile characters in the adult state or vice versa.

**Newcommer's fluid:** A fixative used in chromosome analysis composed of 6 parts isopropyl alcohol:3 parts propionic acid:1 part petroleum ether:1 part acetone:1 part dioxane (with or without ferric acetate).

**NFT:** See **nutrient film technique**.

**niacin or nicotinic acid ($C_6H_5NO_2$ mw 123.11):** Vitamin $B_3$ forming part of a respiratory coenzyme. It is a common micronutrient addition to plant tissue culture media. It is sometimes also known as vitamin PP.

**niacinamide or nicotinamide ($C_6H_6N_2O$ mw 122.12):** An amide of niacin occasionally added to plant tissue culture media.

**nickel (Ni aw 58.70 an 28):** A metallic element sometimes included in plant tissue culture media as **nickel chloride**.

**nicotinamide:** See **niacinamide**.

**nicotinic acid:** See **niacin**.

**nipple flask:** A flat bottomed, round sided flask with many side projections or nipples. Suspension cultures are agitated and aerated in these flasks as medium flows into and out of the projections while the apparatus rotates. The flasks and rotating apparatus were designed by Steward et al. ca. 1952.

**nitrate ion ($NO_3^-$):** A nitric acid ester or salt. A form of nitrogen assimilable by plants and included in virtually all plant tissue culture media in the form of **nitrate** salts.

**nitrogen (N aw 14.0067 an 7):** Present in the atmosphere as $N_2$, a colorless, odorless, inert, gaseous element comprising ca. 78% (volume) of the atmosphere. It is an essential constituent of proteins, nucleic acids, some plant hormones and chlorophyll molecules among others. Its concentration in the fixed forms ($NO_3$, $NH_4$) affects plant growth rates. In excess it supresses fruiting while its deficiency causes stunting and leaf yellowing. It is added to plant tissue culture medium as ammonium ($NH_4^+$) and nitrate ($NO_3^-$) salts.

**nitrogen fixation:** Involves the conversion of atmospheric nitrogen into ammonia, nitrates and other nitrogen-containing compounds by **nitrogen fixing** bacteria, some photosynthetic bacteria and some blue-green algae. The nitrogen fixing bacteria include (free-living) *Clostridium*, *Azotobacter* and (nodulating) *Rhizobium*.

**Nitsch, J.P. (1963):** Described medium supplemented with extract from embryogenic anthers, and induced embryogenesis from isolated microspores of certain plants, such as *Datura*, tobacco and tomato.

**Nitsch, J.P. and C. Nitsch (1969):** Developed a now widely used plant tissue culture medium formulation **NN**, for use in anther culture.

**NN:** An abbreviation for the **Nitsch, J.P. and C. Nitsch (1969)** medium formulation.

**NOA:** See **(2-naphthalenyloxy)acetic acid**.

**Nobécourt, J.R. (1937):** Was among the first to describe the prolonged culture of carrot callus.

**nodal culture:** The culture of nodal explants, usually including a lateral bud and a section of adjacent stem tissue.

**node:** The site (or region, **nodal**) of bud or leaf attachment on a stem.

**nodular:** Describes callus with a pebbly (rough) exterior.

**non-competent:** Cells or tissues incapable of undergoing morphogenesis. The antonym is **competent**.

**nucellar embryony:** The process by which individual cells of the nucellus give rise to somatic embryos (embryoids). Plants on which this occurs naturally are considered good candidates for tissue culture of somatic embryoids.

**nucellus:** A rounded mass of tissue within an ovule, containing the embryo sac and surrounded by the integuments. At fertilization the nucellus may be resorbed as the embryo develops or it may persist, in some seeds, to form a nutritive periderm.

**nuclease:** An enzyme that can hydrolyze internucleotide linkages of a nucleic acid.

**nucleic acid:** Polynucleotide chains which are of fundamental importance to life; they determine the genetic properties of an organism. There are two forms: DNA and RNA; each is composed of long chains of mononucleotides linked by successive 3',5'-phosphodiester bonds (-O-P-O-) between the adjacent ribo- or deoxyribonucleotides of the chain. The nitrogenous base sequence of the nucleic acid chain is maintained by these bonds.

**nucleo-cytoplasmic ratio:** The ratio of cell nuclear to cytoplasmic volume. This ratio is elevated in meristematic cells and low in differentiated cells.

**nucleoside:** A compound formed from a five carbon sugar and a purine or pyrimidine base.

**nucleotide:** A compound formed from a five carbon sugar, phosphoric acid and a purine or pyrimidine base. These compounds consititute part of various coenzymes and building blocks of DNA and RNA.

**nucleus:** **1.** A specialized membrane-bound protoplasmic body inside an eukaryotic cell, containing the genetic material (chromosomes) of the cell and essential to synthetic and developmental cell functions. **2.** The particle about which a mass of associated material can gather or form; as when a crystal accumulates.

**nullisomic:** An abnormal, aneuploid chromosome complement in which both members of a chromosome pair are lacking in the diploid set. Such zygotes are often inviable. The condition is **nullisomy**.

**nurse:** A culture technique or the callus upon which a filter paper is placed separating single cells from the callus in the paper raft technique of Muir (1953). The callus (nurse tissue) releases growth factors and nutrients that induce growth in the single cells supported by the filter paper (those being **nursed**) and sharing the communal environment.

**nutrient:** A nutritive substance or ingredient; as are the major and minor mineral elements (macroelements and microelements) necessary for plant growth and development as well as the organic addenda, sugars, vitamins, amino acids and others employed in plant tissue culture media.

**nutrient film technique (NFT):** Hydroponic plant growth whereby plant roots are suspended in shallow, slowly circulating nutrient solutions delivered as a continuous film of liquid rather than in bursts of liquid.

**nutrient gradient:** A diffusion gradient of nutrients and gases is set up in tissues where only a portion of the tissue is in contact with the medium. This is common in callus culture and causes differential growth rates and senescent regions of callus. Similarly diffusion gradients of nutrients are set up in the medium adjacent to the tissue. Gradients are less likely to form in liquid media.

**nutrient medium or basal nutrient medium:** See **medium formulation**.

# O

**O:** The chemical symbol for the element **oxygen**.

**oat coleoptile test:** A quantitative auxin assay. This determination is based on the degree of curvature, a result of asymetric growth, in a detipped oat coleoptile affixed to agar containing the auxin.

**offset or offshoot:** An asexually produced sucker or basal shoot, bulbil or cormlet.

**ontogeny:** The developmental (**ontogenetic**) life history of an organism, its development from fertilized egg (zygote) to seed-forming adult (reproductive phase), with emphasis on embryonic development.

**open continuous culture:** A cell suspension culture with a continuous influx of fresh medium, maintained at constant volume by the efflux of cells and spent medium.

**organ:** A differentiated, organized plant structure (leaf, root, flower etc.) composed of more than one tissue that performs as a functional and structural unit.

**organ culture:** The growth in aseptic culture of plant organs such as roots or shoots, beginning with organ primordia or segments and maintaining the characteristics of the organ.

**organelle:** A persistent, membrane bound structure with a specialized function in the cytoplasm of cells, such as mitochondria, chloroplasts, plastids, etc.

**organic: 1.** Carbon-containing compound. **2.** Relating to organs or organisms. Antonym of **inorganic**.

**organic cosolvent:** Compound used to dissolve some neutral organic substances; as in media preparation. These include alcohol (usually ethanol), acetone and dimethylsulfoxide (DMSO).

**organism:** A unicellular or multicellular living creature.

**organized growth:** The in vitro development of organized explants such as meristem tips or shoot tips, floral buds or organ primordia, or their de novo formation from **unorganized** tissues.

**organogenesis:** The initiation (de novo) and growth of organs (roots and shoots, usually) from cells or tissues; as in organ culture. Organs may form on the surface of explants (direct organogenesis) or upon an intervening callus phase (indirect organogenesis).

**organoid:** An anomalous organ-like structure formed in culture; as on leaves, roots or callus.

**ortet:** The original mother plant or donor plant from which vegetatively propagated plants are derived.

**orthotropic:** Vertically oriented; with an upright growth habit. The antonym is plagiotropic.

**osmium tetroxide or osmic acid ($OsO_4$ mw 254.20):** A fixing agent used to prepare tissues for electron microscopy.

**osmolarity:** The total molar concentration of the solutes affecting the **osmotic** potential of a solution or nutrient medium.

**osmometer:** A device for measuring the osmotic pressure (water potential) of solutions.

**osmosis:** The diffusion or flow of a solvent, such as water, through a semipermeable membrane into a liquid phase containing solutes in higher concentration, tending to equalize the concentration on each side of the membrane; as water enters through root hairs into a plant.

**osmotic potential or solute potential:** A component of water potential that is always negative, the relative free energy of water, related to the concentration of solute particles (ions or molecules).

**osmotic pressure:** The hydrostatic pressure generated by osmotic flow of a solvent (like water) through a semipermeable membrane into an aqueous phase containing a higher concentration of the solute. Plant cells function as osmometers, which affects water movement between plant cells and in and out of roots. Osmotic pressure is measured with an osmometer.

**osmoticum:** An agent such as glucose or sucrose employed to maintain the nutrient medium osmotic potential equivalent to that of the cultured cells (isotonic). This prevents cell damage in vitro.

**ovary:** The basal, usually expanded, hollow portion of the pistil of a flower, composed of one or more carpels and containing the ovules or seeds; immature fruit. Ovaries are common explant sources for plant tissue culture.

**ovulary culture:** A culture in which the explant is an ovary containing the ovule(s). The ovules may be fertilized in culture (in vitro fertilization). This technique is used primarily when the ovary is essential to proper embryo development.

**ovule:** A structure within a flower ovary that develops into the seed after fertilization. At maturity it is composed of an embryo sac, nucellus, integuments and a stalk. It is a commonly used explant in plant tissue culture.

**ovule culture:** A culture derived from an explanted ovule; which may be fertilized in culture (in vitro fertilization). This technique is used to study development of zygotes and young embryos and is sometimes used to rescue embryos susceptible to abortion when embryo culture is not possible.

**ovum, pl. ova:** A mature female gamete (germ cell or unfertilized egg cell). This is usually a large, immobile, haploid cell with cytoplasm, capable of developing into a new individual of the species and produced by the female ovary or gonad.

**oxidant or oxidizing agent:** A substance that accepts electrons (is reduced) in an oxidation-reduction reaction.

**oxidation:** **1.** The addition of oxygen to a substance with the liberation of heat; as in burning. **2.** The removal of hydrogen or the loss of electrons from a substance.

**oxygen (O aw 15.9994 an 8):** An odorless, invisible, gaseous element occurring as $O_2$ and comprising 21% (volume) of the atmosphere. A component of almost all matter, it readily combines with most other elements. As a macroelement it is essential for plant and animal life. Plants produce $O_2$ during photosynthesis by splitting water.

**oxygenase:** An enzyme enabling an organism or system to utilize atmospheric oxygen.

**ozone ($O_3$ mw 48.00):** A reactive form of oxygen produced by the discharge of electricity in air or oxygen in the atmosphere by the action of solar ultra violet (u.v.) or by u.v. in laboratory sterilizing lamps. It has a faint blue color and faint chlorine-like odor. It is used to sterilize and purify water and air and in bleaching. It is injurious to plants and people at concentrations greater than 2 $\mu$l/l.

# P

**P:** The chemical symbol for the element **phosphorus**.

**$P_1$:** Designates the **parental** generation of a cross initiating a breeding experiment. **$P_2$** is the grandparental, **$P_3$** the greatgrandparental generation, etc.

**packed cell volume (PCV):** A quantitative method of estimating cell growth. It is based on the total cell volume in an aliquot of suspension culture. The aliquot is centrifuged for 5 minutes at 200 g after which the packed cell volume is expressed as a percentage of the aliquot volume.

**palisade:** The parenchyma cells of a leaf, found beneath the adaxial epidermis in mesophytes, or on both sides of the leaf in many xerophytes, bearing chloroplasts and functioning in photosynthesis. These cells are often columnar in shape with their long axis at right angles to the leaf surface. Together with the spongy parenchyma, they form the leaf mesophyll.

**panicle:** A type of inflorescence, in which the main axis has several alternate or spirally-formed branches (racemes) each bearing one or more flowers. Considered the most primitive type of racemose inflorescence.

**panicle culture:** Aseptic culture of grain panicle segments, usually in an effort to induce microspore development.

**pantothenic acid ($C_9H_{17}NO_5$ mw 219.23):** Also known as vitamin $B_5$, it is of wide occurrence in most plant and animal tissue, is essential for cell growth and is an important coenzyme in fat metabolism. It is added to some plant tissue culture media as the calcium salt.

**papain:** A water soluble proteolytic enzyme (protease) extracted from papaya fruit and used especially as a meat tenderizer.

**paper raft technique:** A technique developed by W.H. Muir (1953) to promote development of single cells taken from suspension cultures in which cells are placed onto filter paper squares set on actively growing callus (nurse tissue). Growth factors and nutrients from the callus tissue diffuse through the filter paper, promoting cell growth and development.

**PAR:** See **photosynthetically active radiation**.

**para-amino-benzoic acid (Paba, $C_7H_7NO_2$ mw 137.12):** An occasional addition (vitamin Bx) to plant tissue culture media.

**para-chlorophenoxyacetic acid (pCPA mw 186.53):** A synthetic growth regulator of the auxin type which is sometimes used in plant tissue culture media. It dissolves in base (ca. 1M KOH or NaOH).

**paraffin (wax):** A translucent, white, solid hydrocarbon with a relatively low melting point. It is used for many purposes, including embedding medium used to support tissue for sectioning, in preparation for light microscopy.

**Parafilm:** The brand name for a stretchable, waxed adherent used as a glassware closure or for other purposes in scientific work.

**parahormone:** A substance with hormone-like properties that is not a secretory product; such as ethylene or carbon dioxide.

**Paraplast:** The brand name for a commonly used paraffin-based embedding medium for tissues prepared for microtome section and study in light microscopy.

**parasexual hybridization:** Refers to genetic recombination by means other than through fertilization of germ cells (parasexual) leading to hybrid cells or individuals; as in hybrid cells or plants derived from somatic cell fusion.

**parenchyma:** Thin-walled (with primary cell walls only), vacuolate cells or tissues (**parenchymatous**) alive at maturity and retaining the capacity for renewed cell division and differentiation. This is the most common plant cell type and is regarded as the basic cell type from which all other cell types evolved.

**parthenogenesis:** The development of a new plant from an unfertilized ovule. Such ovules are usually diploid, and the offspring are genetically identical to the parent. **Parthenocarpic** fruit are usually seedless.

**particle radiation:** Refers to $\alpha$-particles (positively charged) and $\beta$-particles (negatively charged), electrons, protons and neutrons. These particles are used to produce mutant cells or organisms in plant tissue culture.

**$\alpha$-particle:** The helium nucleus ($He^{2+}$), emitted at high velocities from certain radioactive disintegrations. See **particle radiation**.

**$\beta$-particle:** A high speed electron (negatively charged) emitted during radioactive decay. See **particle radiation**.

**parts per million (ppm):** This term has largely been replaced by the equivalent mg/l of solution or $\mu$l/l for liquids or gasses.

**passage:** **1.** See **subculture**. **2.** The action, process or means of passing.

**passage number:** The number of subculture intervals. Culture age is a function of passage number and the culture dilution ratio.

**passage time:** The interval between subcultures or the culture period.

**passive absorption:** Ion movement into cells or plant roots in absorbed water. Passive absorption is a plant response to the osmotic concentration gradients, so metabolic energy is not required.

**Pasteur pipette:** A non-calibrated glass or plastic pipette used with a terminal bulb to transfer small liquid volumes. After Louis Pasteur (1822–1895).

**pathogen:** The causal agent of a disease; as a **pathogenic** fungus, bacterium or virus.

**pathogen-free:** A culture or plant that is apparently healthy and protected from infection during propagation and maintenance. Freedom from the specific pathogens for which a plant is being tested and not necessarily a guarantee of good health.

**PBA:** See ***N*-(phenylmethyl)-9-(tetrahydro-2*H*-pyran-2-yl)-9*H*-purin-6-amine** or **6-(benzylamino)-9-(2-tetrahydropyranyl)-9*H*-purine**.

**PCV:** See **packed cell volume**.

**peat:** Incompletely decomposed dead plant material, from waterlogged, anoxic areas. Peat is a commonly used component of horticultural potting mix due to its water-retaining properties.

**pectin:** A high molecular weight, methylated galacturonic acid polymer located in the primary cell walls and in the middle lamella of cells, in the form of magnesium and calcium salts. **Pectinaceous** materials cement cells together.

**pectinase:** An enzyme that degrades **pectin**, the adhesive material that cements cells together. It is used alone or with other enzymes to digest the polygalacturonic acid of plant cell walls (to sugar and galacturonic acid) in protoplast production.

**pedicel or pedicle:** The stem of a flower or flower stalk.

**PEG:** See **polyethylene glycol.**

**penicillin:** A group of closely related bacteriostatic antibiotics, which inhibit gram positive bacteria and some other microorganisms. Penicillin is sometimes included in plant tissue culture media, when other methods to control contamination are not appropriate.

**periblem:** In the histogen theory, the plant apical meristem tissue from which the cortex is derived.

**peptide:** A compound consisting of two or more amino acids covalently linked. **Peptides** with three or more amino acids are **polypeptides**.

**pentose:** A five carbon monosaccharide with the general formula $\mathbf{C_5H_{10}O_5}$, as is ribose, ribulose, xylose and xylulose. Ribose and a deoxy-form, deoxyribose are important subunits of nucleotide or nucleic acid structure. Ribulose in biphosphate form is the $CO_2$ acceptor for $C_3$ photosynthetic carbon assimilation. Xylose, ribose and xylulose play a role in the regeneration of ribulose for carbon assimilation (Calvin cycle).

**periclinal:** The plane of cell division or cell wall orientation parallel to the surface of the organ.

**periclinal chimera:** Refers to a condition in which genotypically or cytoplasmically different tissues are arranged in concentric layers.

**periderm:** A protective tissue comprising the cork, cork cambium and secondary cortex, and replacing the epidermis as the outer layer of the stem and root in plants exhibiting secondary growth. Periderm may also develop at wound sites.

**persistent:** **1.** Continues to exist or to remain attached; as leaves from cultured plants may endure after a plantlet is transferred from culture to soil (persistent leaves). **2.** Chemicals with a long inactivation time, such as some pesticides, which may accumulate in the food chain.

**petiole:** The stalk of the leaf which joins the leaf blade to the stem.

**petri dish (plate):** A shallow plastic or glass covered dish of 50, 90 or 140 mm diameter (most commonly). These are used as culture containers for plant tissue cultures or microorganisms grown on nutrient medium.

**pH:** A measurement of the degree of acidity or alkalinity of a solution. The negative log of the hydrogen ion concentration, in moles per litre (M/l). On the pH scale from 0–14, there is a 10 fold difference for each unit of change of pH. The lower the value, the more acid (more hydrogen ions it contains); pH 7 is neutral; the higher the value, the more alkaline (more hydroxyl ions it contains). Most culture media are adjusted in the range pH 4–6 using 1 M NaOH or HCl.

**pH drift:** A shift in pH that occurs as cultures grow and is related to the buffering capacity of the medium.

**pH meter:** A device for testing the acidity or alkalinity of a solution. An electrode probe is inserted into the liquid sample and a reading taken.

**phellem:** See **cork**.

**phelloderm:** See **secondary cortex**.

**phellogen:** See **cork cambium**.

**phenocopy:** A non-hereditary phenotypic change that is environmentally induced, during a limited developmental phase of an organism, that mimics the effect of a known genetic mutation.

**phenolic oxidation:** The process by which many plant species which contain phenolics blacken through oxidation when they are wounded; as in vitro, soon after explantation or subculture. This may lead to growth inhibition or, in severe cases, to tissue necrosis and death. Antioxidants are incorporated into the sterilizing solution or isolation medium to prevent or reduce oxidative browning.

**phenols or phenolics:** A class of organic compounds with hydroxyl group(s) attached to the benzene ring forming esters, ethers and salts. **Phenolic** substances may bleed from newly explanted tissues, oxidizing to form colored compounds visible in nutrient media or on the filter wicks supporting the tissue.

**phenotype:** The external or visible characteristics of an organism, the expression of its genetic composition (genotype) which sets the reaction norm for environmental influences, and the environment during the organism's development. **Phenotypic** flexibility is the organism's capacity to function in a range of environments. These adaptations may include both plastic and stable responses. The result of interactions between genotype, environment (internal and external) and time.

**_N_-(phenylmethyl)-1_H_-purin-6-amine or 6-benzylaminopurine or 6-benzyladenine (BAP or BA, $C_{12}H_{11}N_5$ mw 225.20):** A synthetic cytokinin commonly employed in plant tissue culture media to induce axillary bud proliferation.

***N*-(phenylmethyl)-9-(tetrahydro-2*H*-pyran-2-yl)-9*H*-purin-6-amine or 6-(benzylamino)-9-(2-tetrahydropyranyl)-9*H*-purine (PBA, $C_{17}H_{19}N_5O$ mw 309.40):** A synthetic cytokinin sometimes used in plant tissue culture media to induce axillary bud proliferation.

**phloem:** The food-conducting tissue, which with xylem tissue forms the vascular tissue of plants. It is a complex tissue composed of sieve elements or sieve cells, companion cells, parenchyma, sclereids and fibres. Primary phloem develops from the procambium of a primary meristem and consists of protophloem and metaphloem. **Protophloem** is the first (primary phloem) to differentiate and become mature. **Metaphloem** matures later. Secondary phloem is derived from the vascular cambium and is composed of several cell types. Sieve cells or elements function in transport of food materials (sugars, organic and amino acids, proteins, etc.). Sieve cells (present in gymnosperms and pteridophytes) are elongated single cells with lateral and terminal sieve areas. Sieve tubes (present in angiosperms) are longitudinal files composed of sieve tube member cells, with one or more sieve areas, forming a sieve plate in the end walls. Each sieve element is closely associated with a companion cell which unlike the sieve element, is nucleate.

**phloretic acid:** A phenolic acid formed by the oxidation of a hydroxyl (OH) of 1,3,5-trihydroxybenzene. An occasional constituent of some plant tissue culture media.

**phloroglucinol or phloroglucin ($C_9H_6O_3$ mw 126.11):** A glucoside of phloretic acid derived from 1,3,5-trihydroxybenzene by oxidation of an -OH to the carboxyl -OOH. Common in some fruit trees such as apple (*Malus* spp.). An occasional constituent of plant tissue culture media. It has some bactericidal properties and stimulates the growth of some cultures.

**phospholipid (phosphoglyceride, phosphatide, glycerophosphatide, phospholipoid):** Any of a group of complex lipids containing a nitrogeneous base and phosphoric acid. The major constituents of cell membranes; as lecithin, phosphatidylethanolamine, etc.

**phosphoric acid ($H_3PO_4$ mw 98):** A strong mineral acid. Phosphate groups ($OPO_4^{2-}$) form important ester links between the subunits of nucleic acids and the two terminal phosphates of di or triphosphonucleotides like ADP and ATP which provide the energy currency for most biochemical processes. Many phosphate bonds are so-called "high energy bonds".

**phosphorus (P aw 30.97376 an 15):** A nonmetallic element present in nucleic acids and involved in chemical energy transfer and many metabolic processes. It is abundant in seeds and fruits. As a macroelement it is added to plant tissue culture media as **potassium phosphate** or **sodium phosphate**.

**photoautotroph:** An organism with the capacity for employing light as the sole energy source in the synthesis of organic materials from inorganic elements or compounds; as do green photosynthesizing plants and some photosynthetic bacteria. The condition is **photoautotrophy**.

**photoheterotroph:** A photosynthetic organism that requires an organic hydrogen source. The condition is **photoheterotrophy**.

**photometer:** An instrument for measuring the luminous intensity of light sources.

**photoperiod:** **1.** The duration of the light portion in an alternating light-dark sequence. This is usually about 16 hours for micropropagated cultures and may be less for some callus cultures. **2.** The optimum light-dark interval for normal plant growth and maturity (evolutionarily determined).

**photoperiodism:** The control of a morphogenetic or physiological response by the length of the **photoperiod**; as in the onset of tuberization, leaf abscision or dormancy. Some responses, such as flowering are controlled by the length of the dark period.

**photorespiration:** Most photosynthetic plants undergo this light-dependent respiration, differing from dark respiration. Its function is unclear. It involves the oxidation of glycollate to glyoxylate in the peroxisomes, followed by recycling to a Calvin cycle carbohydrate. This process wastes $CO_2$, ATP and $NADPH_2$, reducing photosynthetic efficiency.

**photosynthesis:** The process by which plant chloroplasts make carbohydrates from carbon dioxide and water in the presence of chlorophyll using light energy and releasing oxygen. Plant tissue cultures are characterized by impaired **photosynthetic** competence.

**photosynthetically active radiation (PAR):** The photon flux expressed as moles or microeinsteins per meter squared per second ($\mu molm^{-2}s^{-1}$ or $\mu Em^{-2}s^{-1}$) or as watts per meter squared ($Wm^{-2}$) over the wavelength range of 400–700 nm; the part of the light spectrum which is primarily absorbed by plants and used in photosynthesis.

**phototroph:** An organism that can derive energy directly from sunlight to synthesize organic compounds. Refers to photoautotrophs and photoheterotrophs.

**phyllotaxy:** The radial leaf order around the stem of a plant. This is a genetically determined species characteristic.

**physical mutagen:** A mutagen of non-biological or non-chemical origin such as electromagnetic radiation (u.v., x-rays and $\gamma$-rays), particle radiation (electrons, neutrons, protons, $\alpha$-particles and $\beta$-particles), heat or mechanical force.

**physiology:** The study of the biochemical activities and functional processes of organisms, including plants or plant parts.

**phytohormone:** See **hormone**.

**phytokinin:** An obsolete term for cytokinin.

**phytol:** The "tail" of a chlorophyll molecule; a hydrophobic long chain alcohol of four isoprene units (twenty carbon atoms).

**phytotron:** A controlled environment building or chamber for studying plant growth under defined conditions.

**picloram or 4-amino-3,5,6-trichloro-2-pyridinecarboxylic acid (TCP mw 241.46):** A synthetic growth regulator of the auxin type with herbicidal properties. It has been used as a forest defoliant (herbicide). It is used in plant tissue culture media to induce rapid callus proliferation. It dissolves in base (ca. 1M KOH or NaOH).

**pigment:** A colored compound; as chlorophyll in green plants. The amount and type of colored material is referred to as **pigmentation**.

**pinocytotic vesicle:** A cytoplasmic membrane-covered vesicle formed by enclosure of materials in the vicinity of the cell plasmalemma by invagination of the membrane. These vesicles are smaller than plasmolytic vesicles and may be formed during protoplast preparation and isolation.

**pipet or pipette:** A device for measuring and transferring small volumes of liquid. Some are not calibrated, others are calibrated to deliver a fixed volume and some can be adjusted to deliver a range of volumes.

**pistil:** The central female floral structure (organ) composed typically of an ovary, style and stigma.

**pith:** A tissue usually of parenchyma cells occupying the center of the stem or root within the vascular cylinder. Pith is a common explant source for plant tissue culture, especially callus culture.

**plagiotropic:** Refers to a horizontal or prostrate growth habit. The antonym is orthotropic.

**plant growth regulator:** See **growth regulator**.

**plant nutrient:** An essential plant element required in large (macronutrient) or trace (micronutrient) amounts.

**plant tissue culture:** A general term encompassing the in vitro culture of plant cells, tissues, organs and whole plantlets.

**plantlet:** A shoot, sometimes exclusively a rooted shoot, growing in culture or derived from culture.

**plasmalemma or plasma membrane:** The external semipermeable plasma membrane of plant cells that bounds the protoplast and is surrounded by the cell wall to which it is appressed when the cell is turgid.

**plasmid:** A small circular molecule of double stranded DNA occurring naturally in bacteria and yeast. They replicate independently as the host cell proliferates. They often carry vital genes such as may confer resistance to environmental, biological or chemical factors. While some plasmids have large and difficult to isolate DNA, others are small in size and can easily be separated from the host cell, genetically modified via gene insertion, and proliferated in a bacterial (usually) cell culture. By means of an appropriate vector (for plants this could be *Agrobacterium tumefaciens*), the modified plasmid genome (or part of it) can be introduced into individual plant cells or protoplasts in vitro. The cells containing the new genetic information can be used to regenerate plants possessing the desired genetic information. New plant cultivars possessing novel characteristics such as resistance to pathogens, herbicides and environmental stress; morphological or physiological characteristics are theoretically possible by these means.

**plasmodesma, pl. plasmodesmata:** Fine cytoplasmic passages, connecting adjacent living plant cells, through which materials can pass. These occur in variable numbers between cells and are sometimes concentrated in primary pit fields (cavities in the secondary cell wall). Each plasmodesma is lined by the plasmalemma and is associated at either end with the endoplasmic reticulum.

**plasmolysis:** The loss of water from a plant cell causing the cytoplasm to separate or shrink from the cell wall. This condition results from placing cells in hypertonic solutions.

**plasmolytic vesicle:** A membrane-bound vesicle present inside the cell cytoplasm, formed from an invagination or infolding of the plasmalemma. They may be several micrometres in diameter and are commonly formed during cell wall removal in protoplast preparation.

**plasmotype:** A cell type displaying features which are expressions of cytoplasmic, rather than nuclear inheritance.

**plasticity:** The range of environmentally-inducible phenotypic expressions.

**plastid:** A small, double membrane-bound, self-duplicating organelle in the cells of eukaryotes, functioning in synthesis or storage mostly of soluble and insoluble carbohydrates. In higher plants these differentiate from proplastids in meristematic regions. These may contain DNA, pigments (chlorophyll and carotenoids) and/or reserve food such as starch, oil or protein. This group of organelles includes **chloroplasts** which contain chlorophyll and photosynthesize; **chromoplasts** which contain carotenoids; **leucoplasts** which have no pigments and store starch; as well as others.

**plate:** **1.** To distribute a thin film of; as microorganisms or plant cells are **plated** onto nutrient agar. In the Bergmann plating technique cells are mixed with liquid nutrient medium just prior to dispensing the medium into a petri dish. **Plating** single cells allows selection of homogeneous single cell lines (colonies) from cell suspensions composed of heterogeneous cells. **2.** Refers to the two segments of a petri dish or a similar-shaped item. Also see **cell plate**.

**platform shaker:** See **shaker**.

**plating efficiency:** An estimate of the percentage of viable cell colonies developing on an agar plate relative to the total number of cells spread onto the plate. Plating efficiency is a function of the tissue (species, explant and cell line); medium composition; plating density and the phase of the stock culture.

**pleiotropy:** The condition in which several characteristics are affected by a single gene.

**plerome:** In the histogen theory, the plant apical meristem tissue which differentiates into all the primary tissues internal to the cortex.

**ploidy:** The number of complete sets of chromosomes possessed by a cell or organism, where the usual number is **diploid** (2n). This number may be affected in culture leading to multiplication (**polyploidy**).

**plumule:** The embryonic axis above the cotyledonary node.

**polar nuclei:** In the center of a developing embryo sac, in the angiosperms, are two haploid nuclei that, when fertilized, develop into triploid endosperm.

**polarity:** **1.** Morphological or physiological dissimilarity between extreme ends of an axis of an organism or structure, or the cells within it; as the shoot and root of a plant are dissimilar. **2.** The tendency of some transported substances to move in one direction within a plant; as IAA moves basipetally in the oat coleoptile.

**pollen or pollen grain:** The partly developed male gametophyte, a germinated microspore. These male spores often appear as yellow dust. They are a common explant source for plant tissue culture, usually aimed at the production of monoploid plants.

**pollen culture:** The culture of pollen grains, which germinate in vitro. Such cultures may eventually form monoploid callus, from which shoots or embryoids develop into monoploid plants.

**pollen sac:** A sac within the anther of a stamen within which microspores (pollen) are produced.

**pollen tube:** An outgrowth of the pollen grain wall that emerges on germination and grows towards the egg carrying the male gametophytes within. In angiosperms the pollen tube grows down the stylar canal, discharging its contents through the micropyle or chalaza. Pollination ensues. The pollen tube contains three nuclei, the tube (vegetative) nucleus and two generative nuclei. One generative nucleus fuses with the egg cell, forming the zygote. In some angiosperms the other generative nucleus fuses with the polar nucleus forming the endosperm nucleus which develops into endosperm.

**pollinate:** To transfer pollen from the anther to the receptive surfaces of the stigma of the same or another flower in angiosperms and from male to female in gymnosperms. This process is **pollination** and usually requires a vector in outbreeding plants. Pollination can be done in culture (in vitro pollination).

**polyamine:** A compound with two or more amino groups; as putrescine ($NH_2(CH_2)_4NH_2$) and spermine ($NH_2(CH_2)_3NH(CH_2)_4NH(CH_2)_3NH_2$).

**polyembryony:** A condition in which more than one **embryo** is formed per ovule by division of the zygote (cleavage polyembryony) or from somatic tissue (adventive polyembryony). In some plant species individual seeds may contain more than one embryo, one of zygotic and the rest of somatic origin.

**polyethylene glycol (PEG) or carbowax:** A fusion-inducing agent (fusogen) for agglutinating protoplasts which is used in somatic hybridization studies. This compound is also sometimes used in media as a non-metabolite osmoticum and is available in various molecular weights, ranging from ca. 200 to 6,000.

**polygalacturonic acid:** A long chain sugar acid polymer composed of galacturonic acid and hexose subunits.

**polygenes:** Systems of genes associated with quantitative character variation in which each gene individually effects the phenotype in a minor way.

**polynucleotide:** A **nucleotide** sequence of several to many units that is covalently linked; the 3′ position of the pentose of one nucleotide is joined to the 5′ position of the pentose of the next one by a phosphate group.

**polyploid:** Genetic state resulting from multiplication of the basic monoploid chromosome set at least three times. The condition is **polyploidy** and is common in plants (triploidy (3n), tetraploidy (4n), pentaploidy (5n), etc.). Polyploid organisms crossed with diploids are sterile, but can reproduce by self fertilization, parthenogenesis or vegetatively. **Polyploidy** is a common occurrence in protoplast, cell, callus and other cultures via cell fusion, asynchrony in nuclear and cytoplasmic division or by other means. It may also be induced by treating cells with certain chemicals such as colchicine.

**polypropylene:** A strong, flexible, transparent thermoplastic formed by the polymerization of propylene. It is used in many labware products.

**polysaccharide:** A chain of one or more monosaccharide (simple sugar) units, or their derivatives, linked by glycosidic bonds; linear or branched. The most abundant **polysaccharides** are cellulose and starch. Cellulose is the structural component of fibrous and woody plant tissues. Starch is the major form of energy storage in plants.

**polyvinylpyrrolidone (PVP, $(C_6H_9NO)n$ mw (111.145)n):** An occasional constituent of plant tissue culture isolation media. It is of variable molecular weight and has antioxidant properties, so is used to prevent oxidative browning of explanted tissues. It is less frequently used as an osmoticum in culture media.

**population density:** Cell number per unit medium area or medium volume.

**porphyrin:** Any of a class of important, naturally occurring pigments derived from porphin, a ring structure of four pyrrole nuclei, linked by carbon atoms; including the chlorophylls, cytochromes and haemochromes.

**post-zygotic incompatibility:** A condition where, in the case of incompatible or wide crosses the zygote fails to develop, often for nutritional reasons. In some cases embryo culture can be used to rescue such embryos.

**potassium (K aw 39.098 an 19):** A reactive, alkaline, metallic element. Potassium is involved in cell salt and water balance; chlorophyll, carbohydrate and protein synthesis; nitrate reduction; carbon dioxide fixation; and normal cell division. It is an important macroelement constituent of plant tissue culture media and is added as **potassium bromide, potassium chloride, potassium iodide, potassium nitrate** or **potassium phosphate**.

**potassium hydroxide (KOH mw 56.11):** A white, solid compound that forms an alkaline caustic solution in water. It is used to adjust the pH of plant tissue culture media and to dissolve auxins.

**potassium hypochlorite (KClO mw 90.6):** A water soluble bleaching or disinfecting agent used in plant tissue culture practices.

**potato extract:** A common (undefined) organic addendum to plant tissue culture media in monocot and anther culture systems.

**ppm:** See **parts per million**.

**preadaptation:** In possession of advanageous characters enabling an organism to be well suited (**adapted**) to environmental conditions previously unknown to it.

**precipitate:** The separation of solid matter from a liquid which usually falls to the bottom of the vessel; as particulate material appears during incubation of tissues in some liquid nutrient media. The process is **precipitation**.

**precipitin test or microprecipitin test:** A serological assay in which visible particulates (**precipitates**) form when soluble antigen and antibody react. This precipitin reaction detects and identifies antigens.

**precondition:** To put into desired **condition** in advance; as some pretreatments such as etiolation or thermotherapy are useful in donor plants or explanted material prior to plant tissue culture initiation to enhance growth in culture.

**prefilter:** A coarse filter (furnace filter) such as those used in a laminar air flow cabinet to screen out large particles before air is forced through a much finer filter (HEPA filter).

**premix:** To mix before use; as nutrient mixtures are commercially available as dry powders of preweighed ingredients, which are put into solution when required.

**preserve:** **1.** To keep safe, guard or protect. **2.** To maintain. **3.** To keep from decaying. The process is **preservation**.

**pressure cooker or canner:** A small-scale alternative to an autoclave working on the same principles. **Pressure cookers** are used for sterilizing small amounts of media or labware.

**pressure potential or turgor potential:** The pressure which can be generated within a cell which is the difference between the water potential of the external environment and the osmotic potential within the cell, provided the cell volume is constant.

**presumptive hybrid:** A cell or individual plant for which there is evidence to suggest that it is a hybrid; as somatic cell fusion products intermediate in color to that of (stained) parental cell lines.

**presumptive or putative mutant:** A cell or individual plant for which there is evidence to suggest that a mutation has taken place; as in some cell variants possibly induced by mutagens, with features not possessed by other cells.

**pretransplant:** Stage III in tissue culture micropropagation; the rooting, hardening stage prior to transfer to soil (Stage IV). Media for this stage usually contain auxins, of which the most common is 1*H*-indole-3-butanoic acid (IBA). Hardening treatments may include increased light intensity; a dark interval; a cold treatment; etc.

**prezygotic incompatibility:** In the case of incompatible or wide crosses the inhibition of pollen germination or the prevention of pollen tube growth, among other possible barriers to plant fertilization. In some cases these barriers can be overcome through in vitro pollination and/or embryo culture.

**primary cell wall:** The cell wall layer formed during cell expansion. Plant cells possessing only primary walls may divide or undergo differentiation.

**primary culture:** **1.** May be synonymous with Stage I culture. **2.** A recent culture not yet subcultured for the first time.

**primary growth:** **1.** Refers to apical meristem-derived growth; the tissues of a young plant. **2.** Refers to explant growth during the initial culture period; as primary callus growth.

**primocane:** The first year, vegetative canes of bramble plants. In the following year the 2 year old canes flower and set fruit, so are known as floricanes, and a new set of **primocanes** develop.

**primordium:** The earliest, rudimentary developmental stage of an organ or cell; as leaf primordia which are barely visible on an excised shoot tip.

**PRL-4:** See **Gamborg, O.L. (1966)**.

**probe DNA:** A radioactively labeled (usually $^{32}P$) DNA molecule used to detect complementary-sequence nucleic acid molecules by molecular hybridization. To localize the probe DNA sequence and reveal the complementary hybridization sequence autoradiography is often used.

**proembryo:** A rudimentary embryo, before the first distinct globular-stage **embryo** or embryoid forms.

**progeny:** The offspring derived from either sexual crosses, or asexual processes.

**prokaryote or procaryote:** Cell or organism in which the nuclear material is not organized into chromosomes nor separated from the cytoplasm by an enclosing membrane; as in bacteria and blue-green algae.

**proliferate:** To grow, multiply or increase by rapid production of new units (cells, plants, etc.). The process is **proliferation**. The state or condition is referred to as **proliferative**.

**promeristem or protomeristem:** The embryonic meristem-containing organ initials or foundation cells.

**propagate:** To reproduce or cause to multiply or breed. The process is **propagation**. Plants are propagated from spores, seeds, cuttings or other propagules.

**propagule:** The form or portion of an organism used for reproduction or propagation; as new shoots or callus derived from explants are subdivided into **propagules** and recultured for further multiplication.

**prophase:** The first phase in mitotic nuclear division, in which chromatin condenses into chromosomes composed of two chromatids, joined at the centromere, the nuclear membrane disintegrates, chromatids migrate to opposite poles and spindle fibers form and attach to the chromosomes.

**propylene oxide ($C_3H_6O$ mw 58.08):** A colorless, flammable, volatile liquid used as a chemical sterilant for plant material and soil, or as a solvent for resins during the preparation of plant material for embedding (light or electron microscopy).

**protease:** A group of water soluble protein-degrading enzymes (proteolytic) that function by breaking peptide linkages. In this group are pepsin (principal protease in the gastric juice of vertebrates), trypsin (protease produced in the pancreas) and papain (a thiol protease derived from papaya fruit).

**protein:** A class of high molecular weight organic compounds with complex combinations of amino acids containing carbon, hydrogen, nitrogen, oxygen and sometimes other elements; present in all living matter. One protein molecule is made up of hundreds or thousands of amino acids, of which about 20 are essential. These are linked by peptide bonds. The sequence of amino acids gives the protein its specific properties. The three dimentional structure (conformation) of proteins determines their biological proportions. Some are globular and are mostly enzymatic in function. Others are fibrous and are usually structural or contractile in function.

**protein digest:** The enzymatic hydrolysis of proteins to yield their building block components, amino acids and short chain peptides (chains consisting of two to several amino acids normaly without enzymatic function).

**Protista:** A major taxonomic group (kingdom) composed originally of all unicellular organisms. Now comprising all simple biological organisms, whether unicellular, coenocytic, or multicellular, including algae, bacteria, fungi, slime molds and protozoa. All are characterized by a true nucleus, chromosomes and usually unicellular reproductive structures.

**protoclone:** Distinct phenotypic regenerants from a plant protoplast. A clone initiated from a protoplast or protoplast-fusion product.

**protocol:** A sequence of activities, techniques or procedures linked to achieve a specific goal; as a schedule for protoplast preparation.

**protocorm:** A tuberous structure formed by orchid seed germination from which orchid plants develop. Similar structures may develop in vitro from somatic explants and can be multiplied or induced to regenerate plantlets.

**protoderm:** In the apical meristem, derivatives of this outermost cell layer give rise to the epidermis and sometimes associated subepidermal tissues of mature structures.

**protomeristem:** See **promeristem**.

**protophloem:** See **phloem**.

**protoplasm:** The viscid, living cell contents or substance of the cell protoplast including the nucleus and the plasma membrane.

**protoplast:** The entire contents of a cell (protoplasmic and otherwise) bounded by the plasmalemma, excepting the cellulosic cell wall. These are prepared by mechanical (limited application) or enzymatic removal of plant cell walls in the presence of an osmoticum.

**protoplast culture:** The isolation and culture of plant protoplasts by mechanical means or by enzymatic digestion of plant tissues or organs, or cultures derived from these. Protoplasts are utilized for selection or hybridization at the cellular level and for a variety of other purposes.

**protoplast fusion:** The coalescence of the plasmalemma and cytoplasm of two or more **protoplasts** in contact with one another. Initial adhesion is a random process but coalescence may be promoted in various ways (induced fusion). When adhesion occurs between adjacent protoplasts during enzymatic wall degradation or between freshly isolated protoplasts in the absence of a fusion agent, it is termed spontaneous fusion.

**proximate or proximal:** Located close to or towards the attachment point of an organ.

**protoxylem:** See **xylem**.

**pseudobulbil:** A thickened bulb-like aerial stem; as possessed by some orchids.

**Pteridophyta:** A plant division containing the nonseed-bearing vascular plants; club mosses, ferns, horsetails, etc. All show heteromorphic alternation of generations. Often the gametophyte is nutritionally independent of the sporophyte. Some species are homosporous, others heterosporous. The sporophyte lacks vessels in the xylem, unlike seed-bearing plants.

**pubescent:** Hairy, as the leaves or other plant organs may be covered with hairs or trichomes.

**pure culture:** Axenic culture.

**pure line:** All cell or individual members are homozygous for one or more characters or genes and will give rise to more cells or organisms with the character(s) under consideration.

**purine:** A nitrogen-containing organic base, with a double ring structure **($C_5H_4N_4$)**, synthesized mainly from amino acids. Two purines, adenine and guanine are commonly present in nucleotides and nucleic acids.

**PVP:** See **polyvinylpyrrolidone**.

**Pyrex:** A brand name for borosilicate glass. Pyrex is a high temperature (autoclavable), impact, abrasion and chemical resistant, stable glass of which much laboratory glassware is made.

**pyridoxine ($B_6$, $C_8H_{11}NO_3$ mw 168.98):** A water soluble, B group vitamin important as a coenzyme (**pyridoxal phosphate, pyridoxanine**) in many metabolic reactions. It is a common addition to plant tissue culture media as **pyridoxine hydrochloride ($C_8H_{11}NO_3.HCl$ mw 205.64)**.

**pyrimidine:** A nitrogen-containing organic base, with a single six-membered heterocyclic ring structure, ($C_4H_4N_2$), synthesized from amino acids. Derivatives are of great biological import; as cytosine, thymine and uracil which occur in nucleotides and nucleic acids.

# Q

**quarantine:** The control of plant import and export for purposes of disease and pest control, and to prevent spread of disease and pests. Control may involve isolation of plant material, sometimes for extended periods of time, to determine its disease status.

**quiescent:** Quiet, at rest, not necessarily dormant and having the potential for resumed activity; can apply to cells unlikely to divide, the non-meristematic cells.

# R

**$R_1$:** Designates **regenerant** from in vitro culture; as a shoot from callus or a plantlet from a somatic embryoid. **$R_2$**, **$R_3$** are the second and third generations from crosses between the **$R_1$**, **$R_2$** generations respectively.

**radiant energy:** Electromagnetic wave energy; as radio waves, gamma, ultraviolet and visible light rays.

**radiation:** Rays of heat, light or particles in wave form.

**radicle:** The embryonic root. It is the first structure to appear when a seed germinates.

**ramet:** The vegetatively propagated offspring of an ortet. An individual of a clone.

**reaction norm:** The range of phenotypic responses of a particular genotype in response to the environmental influences.

**reagent:** A chemical substance that participates in or produces a chemical reaction.

**recessive:** Designating a gene (recessive gene) that, appearing in a hybrid offspring, is suppressed by the dominant gene and expressed under homozygous not heterozygous allele conditions. The antonym of dominant.

**reciprocating shaker:** A platform shaker with a back and forth action used for agitating culture flasks at variable speeds.

**recombinant DNA:** Genetic material with novel gene sequences produced by cross-overs, chromosome reassortment, by other natural means or through genetic engineering.

**recombinant DNA technology:** See **genetic engineering**.

**recombination:** The appearance in offspring of new trait (gene) combinations not exhibited by the parents. Recombination occurs through crossing over and chromosome reassortment during meiosis, followed by random union of different sorts of gametes at fertilization, and is a major source of variation.

**reconstructed cell or recon:** A viable cell hybrid, cybrid or a transformed cell resulting from genetic engineering.

**reculture or subculture:** The aseptic transfer of a culture or a portion (division when necessary) of that culture (inoculum) to fresh nutrient medium.

**redifferentiation:** Cell or tissue reversal in differentiation from one type to another type of cell or tissue.

**redox:** Oxidizing and reducing reactions involving transfer of an electron from a donor to an acceptor molecule; the donor becomes oxidized and the acceptor is reduced. Arranged in order of intensity these reactions are known as the redox series.

**reduced nitrogen:** Nitrogen in the form of $NH_2$ (amine group); $NH_3$ (ammonia) or $NH_4^+$ (ammonium) ion.

**regenerate:** **1.** To form or create again; reform. **2.** To give or gain new life or to renew (heal) by a new growth of tissue; as **regenerants** (shoots, plantlets) form from explants in plant tissue cultures. The process is **regeneration**.

**regime or regimen:** A systematic treatment.

**Reinert, J. (1958):** Among the first to observe adventive embryogenesis in cell cultures of carrot.

**rejuvenation:** **1.** Synonymous with dedifferentiation. **2.** Treatment that leads to culture invigoration (such as subculture) or revival (dormancy breaking).

**relative humidity (RH):** The percentage of ambient moisture vapor compared with that in a saturated atmosphere at any given temperature. Relative humidity is very high (close to saturation) in vitro. Acclimatization to ex vitro conditions includes adaptation to lower relative humidity environments.

**Repelcote:** The brand name of a material (dimethyl dichloro-silane) used to coat glassware (soda glass); for tissue culture containers and for other purposes.

**replicate:** **1.** To duplicate. **2.** To repeat an experiment or procedure.

**replication:** **1.** The act or process of creating a clone. **2.** The mechanism by which DNA strands duplicate at each cell division, using themselves as templates, is termed semiconservative replication.

**repression:** Altered gene expression resulting in the failure of a specific protein synthesis.

**reproduce:** To create another individual of the parental type, that will in turn produce another. The process is **reproduction**.

**reproduction:** A facsimile of the original; as a new individual is formed by sexual or asexual means.

**resistance:** **1.** The ability of an organism to withstand stress, the force or effect of disease, disease agents or toxic substances. **2.** The ability of an organism to resist change due to environmental or other factors. **3.** Impedence to the flow of electrical current.

**resistance transfer factor:** A plasmid, present in certain types of bacteria such as *E. coli*, that can impart resistance to antibiotics in animals exposed to them.

**resistivity:** The degree of resistance (in ohm-cm or ohm-m) or interference to the flow of an electrical current or the movement of particles from place to place. A measure of water purity; the purer the water, the greater its **resistivity**. The reciprocal of conductivity.

**respiration:** **1.** Gas exchange; as by breathing, oxygen uptake, water and carbon dioxide release. **2.** The process involved in liberation of energy and its use within a cell; as in sugar breakdown through oxidation.

**response:** A change in an organism or its parts induced by a stimulus.

**reverse osmosis:** A procedure in which a solution is forced through a membrane against an osmotic gradient, filtering out impurities in the solution. This procedure, deionization or distillation, is a first step in water purification.

**rhizogenesis:** Root formation and growth; as in root development de novo from callus.

**rhizome:** A usually underground, horizontal stem, the primary shoot of the plant. It usually has buds in the axils or reduced leaves and functions in vegetative propagation. Sometimes **rhizomes** are thickened and contain stored food. These may act as overwintering organs.

**rhodamine isothiocyanate or tetramethylrhodamine isothiocyanate ($C_{25}N_3SO_3H_{21}$ mw 536.10):** A dye used to stain plant protoplasts in the identification of fusion products.

**ribavirin:** See **virazole**.

**rib meristem or file meristem:** A meristematic tissue producing parallel files of cells.

**riboflavin or lactoflavin ($C_{17}H_{20}N_4O_6$ mw 376.36):** A water soluble, vitamin ($B_2$) essential to cellular respiration. It is important in carbohydrate metabolism and implicated in photooxidation of auxins and perception of phototrophic stimuli. Riboflavin is an occasional addition to plant tissue culture media. It is sometimes called vitamin G.

**1-β-D-ribofuranosyl-1,2,4-triazole-3-carboxamide:** See **virazole**.

**9-β-D-ribofuranosylzeatin:** A natural hormone (cytokinin) present in coconut milk.

**ribonucleic acid (RNA):** A polynucleotide present in ribosomes and occurring as strands of messenger RNA and transfer RNA. These strands translate the DNA coding sequence into protein (amino acid) sequences.

**ribonucleic acid, messenger (mRNA):** A ribonucleotidal chain produced within the cell nucleus which codes for a specific protein synthesized on ribosomes in the cytoplasm.

**ribonucleic acid, transfer (tRNA):** A class of RNA molecules that covalently bond to specific amino acids. These in turn hydrogen-bond to an mRNA triplet or codon for that amino acid.

**ribose ($C_5H_{10}O_5$ mw 150.13):** A five carbon (pentose) sugar, component of riboflavin, ribonucleic acid (RNA) and cytokinins. An occasional addition to plant tissue culture media.

**ribosome:** Cytoplasmic cell particle of equal parts protein and RNA, present in all organisms and associated with the endoplasmic reticulum in eukaryotic cells. **Ribosomes** are the sites of protein synthesis. Messenger RNA (mRNA) attaches to them and receives transfer RNA molecules bearing amino acids. The developing polypeptide chain and the ribosome are translocated along the mRNA molecule during translation of the genetic code in protein synthesis.

**ribulose bisphosphate (RuBP):** Di-phosphorylated ribulose, which accepts carbon dioxide as the first step in the Calvin cycle of photosynthesis.

**ribulose bisphosphate carboxylase (RuBPase) or carboxydismutase:** The main protein component of chloroplasts and a key photosynthetic enzyme which catalyzes the addition of carbon dioxide to ribulose bisphosphate. It also acts as an oxygenase which catalyzes the addition of oxygen to ribulose bisphosphate, a key step in photorespiration.

**RNA:** See **ribonucleic acid**.

**mRNA:** See **ribonucleic acid, messenger**.

**tRNA:** See **ribonucleic acid, transfer**.

**rogue:** **1.** To critically evaluate and eliminate unwanted plants from a population; as undesirable phenotypic variants, weeds and diseased plants are destroyed in plant propagation practices. **2.** A variant plant in a population (sport).

**Roentgen or Rontgen ray:** See **x-ray**.

**root:** Part of the descending vascular plant axis, generally below the ground and serving to anchor the plant; to supply it with water and minerals and in some species to store food.

**root culture:** Isolated root tips of apical or lateral origin may produce in vitro root systems with indeterminate growth habits. These were among the first kinds of plant tissue cultures (White, 1934) and remain important research tools in the study of developmental phenomena; mycorrhizal; symbiotic and plant-parasitic relationships.

**root hair:** A short-lived, tubular, microscopic, epidermal cell outgrowth found in the root maturation zone. They function by absorbing water and minerals from the soil. In vitro plantlets often lack or have relatively few root hairs on solid medium.

**rooting:** The process of initiating and developing **roots**.

**rootstock:** The upright underground stem part of a plant; as the portion to which a scion is attached in grafting.

**rotary shaker:** A platform shaker with a circular motion used for agitating culture flasks at variable speeds.

**rotating culture or rotator:** A wheel-like device for slowly (ca. 1 rpm) rotating and gently agitating cultures, usually in a vertical plane.

**RuBP:** See **ribulose bisphosphate**.

**RuBPase:** See **ribulose bisphosphate carboxylase**.

**rudiment:** An organ or structure in its initial developmental stage (primordium) or one in which development has been arrested (vestigial).

**runner:** A slim horizontal branch that roots at the nodes and near the tip, giving rise to daughter plants; as the strawberry stolon.

# S

**S:** The chemical symbol for the element **sulphur**.

**saccharide:** See **sugar**.

**saccharose:** See **sucrose**.

**sal amoniac:** See **ammonium chloride**.

**salt:** **1.** Sodium chloride (NaCl mw 58.44). **2.** A compound (along with water) resulting from the reaction of an acid and a base in which the hydrogen of the acid is replaced by a metal from the base.

**salt tolerance or saline resistance:** The ability to tolerate or resist (withstand) a concentration of sodium ($Na^+$ ion) in the soil (or in culture) which is damaging or lethal to other plants. Sodium toxicity is a global problem on irrigated lands. Breeding and screening for increased tolerance and resistance in agronomic plants is of great current interest.

**sanitation:** The process or the result of disinfection of tools, working areas, plant materials, hands, etc.; as for plant tissue culture manipulations. Culture incubation areas must be **sanitary**.

**scaling:** The removal and planting of bulb scales to induce rooting is commonly employed to propagate bulb crops. Bulblet production may be enhanced by scoring (making cuts across the basal plate) or scooping (scooping out the basal plate).

**scalpel:** A straight knife which can have a disposable thin blade. They are used in explant and subculture manipulations by plant tissue culturists.

**scanning electron microscope (SEM):** A microscope used to examine the surface structure of (specially prepared) biological specimens. A three dimentional screen image is acquired through focusing secondary electrons emitted from a sample surface bombarded by an electron beam.

**Schenk, R.U. and A.C. Hildebrandt (1972):** Developed a medium formulation (**SH**) widely used in plant tissue culture, for induction and culture of callus in a wide range of both monocot and dicot species.

**Schiff's reagent:** A mixture of pararosaniline.HCl (an analine dye) and sodium bisulfite ($NaHSO_3$), used in staining chromosomes and nuclear material. See **Feulgen's test**.

**Schleiden, M.V. and T. Schwann (1838):** German biologists who proposed the cell theory, in an attempt to explain the complexity of multicellular organisms. They suggested that each cell of an organism has the potential to recreate an entire new organism and that differentiated cells retain this capacity, through retention of information present in the zygote.

**scion or cion:** The portion of a bud or shoot used for grafting onto another plant (the stock).

**sclerenchyma:** A tissue composed of cells (fibres, sclereids) with thick secondary walls which are often heavily lignified. These cells provide a support function and are often non-living at maturity. Fibres are elongate with tapered ends, occurring singly or in strands. Sclereids are generally squarish to rectangular or irregularly shaped (astrosclereids) and occur singly or in small groups.

**screen:** To separate by exclusion or collection on the basis of a set of criteria (biochemical, anatomical, physiological, etc.); as on the basis of size by passing through a mesh of a specific size or as in excluding the sun's rays. The process is **screening**. This term is often applied to the process of plant selection for disease resistance or improved horticultural qualities.

**screen house:** A metal or plastic mesh enclosure similar in function to a greenhouse but with more exposure to the elements.

**scutellum:** That part of the cotyledon in Gramineae (grasses) that absorbs food from the endosperm at germination. An explant source for plant tissue culture.

**secondary:** Following the first in development or in importance.

**secondary cell wall:** A structure of cellulose and other materials (lignins) inside of and adjacent to the primary cell wall. It functions in support.

**secondary cortex or phelloderm:** The inner layer of periderm, produced toward the inside by cork cambium (phellogen) and consisting of cells that resemble cortical parenchyma.

**secondary meristem:** A meristem producing secondary wall thickening or secondary plant growth.

**secondary product synthesis:** The often large scale utilization of plant cells for industrial production of specific plant products such as gums, resins, alkaloids, flavorings, enzymes or pharmaceuticals.

**secondary thickening:** Cambial activity giving rise to additional (secondary) vascular tissue (conduction and support tissue) resulting in increased stem or root girth.

**sector:** An explant or subculture portion consisting of a cutting from the main root axis bearing four or five young lateral roots.

**seed:** Formed after fertilization by maturation of the ovule of seed plants or in some plants by agamospermy. Gymnosperm **seeds** have an embryo, a seed coat and haploid storage tissue. Angiosperm seeds have an embryo, a seed coat and may have triploid endosperm (storage tissue). Seeds may be surface sterilized and germinated in vitro and are frequently used as a source of aseptic explant material.

**seed leaf:** See **cotyledon**.

**seedling:** **1.** A plant grown from **seed**. **2.** A young plant.

**selection:** **1.** The identification or separation from a group; at the cellular or plant level for tolerance, resistance or some other special quality. **2.** Among organisms (natural selection) and may have evolutionary connotations. **3.** A new hybrid or unique plant that is a candidate for formal cultivar release by the plant breeder.

**selection culture:** Utilizes difference(s) in environmental conditions or more usually in culture medium composition, such that preferred variant cells or cell lines (presumptive or putative mutants) are favored over other variants or the wild-type.

**selection pressure:** The measure of the effectiveness of natural or experimental selection in altering the genetic composition of a population.

**selection unit:** Single cells or small clusters, units of optimum size for isolating and regenerating variants or mutants; the minimum number of cells effective in the screening process.

**selective advantage:** Implies the possession of increased fitness within an individual or a population.

**selective agent:** An environmental or chemical agent that imposes a lethal or sublethal stress on growing plants, or portion thereof in culture, enabling selection of resistant or tolerant individuals.

**self:** To cross an individual plant with itself or with another plant of the same cell line; as intergeneric somatic hybrids may be interbred or self pollinated. The process is **selfing**.

**SEM:** See **scanning electron microscope**.

**semiconservative replication:** See **replication**.

**semicontinuous culture:** The maintenance of cells in a culture vessel in an actively dividing state by periodically draining the medium and adding fresh medium.

**seminal leaf:** See **cotyledon**.

**semipermeable membrane:** A cell or plasma membrane that is partially permeable; certain ions or molecules can pass through it but others cannot. **Semipermeable membranes** will allow the passage of water and solvents, but not certain solutes.

**semisolid:** Gelled but not firmly so; as are media made with small amounts of a gelling agent.

**senescence:** The state, or process of becoming old or aging; as a **senescent** organ or plant gradually deteriorates. This is part of the normal developmental sequence of plants and is genetically determined.

**sepal:** A modified leaf. The **sepals** form the calyx of a flower, and protect the flower bud. In some plants these are petalloid and attract pollinating insects.

**sequestration:** The process of complexing metal ions, usually to chelating agents, to render them ineffective.

**Sequestrene 330 Fe:** The brand name for an iron chelate used in some tissue culture media.

**serial float culture:** Sunderland's (1977–1979) technique of floating anthers on liquid medium and subculturing them to new medium at several day intervals as anther dehiscence, pollen release and development occur, increasing anther productivity.

**serial section:** Series of consecutive tissue sections, usually made with a microtome.

**serine (Ser, $C_3H_7NO_3$ mw 105.09):** An amino acid occasionally added to plant tissue culture media as a source of reduced nitrogen.

**serology:** The study of serum reactions; those between an antigen and its antibody. These tests are used to identify and distinguish between antigens; as specific microorganisms or viruses. They are employed as indicators to assay plants suspected of being virus-infected.

**serologically specific electron microscopy (SSEM):** Grids for an electron microscope are pre-coated with a specific antiserum then floated on tissue samples (macerates). Specific virus particles become attached to the grid. SSEM is rapid, reliable, more sensitive than many other serological tests and uses less antiserum.

**serum:** The fluid fraction of vertebrate blood remaining after the red cells have been removed. It provides the basis for antigen-antibody reaction tests. See **serology**.

**serum cap:** A rubber, gas impermeable tube closure.

**sexual reproduction:** The fusion of haploid nuclei, usually of gametes; followed by meiosis then recombination. This process occurs during the reproductive (seed forming) phase in the life of an organism. Special types of sexual reproduction include apomixis and parthenogenesis which involve development of individuals from female gametes without the occurrence of fusion. The antonym is asexual reproduction.

**shake culture:** An agitated suspension culture. Usually a flask (commonly an Erlenmeyer flask) containing the culture is attached to a horizontal or platform shaker, or agitated with a magnetic stirrer, to provide adequate aeration for cells in the liquid medium.

**shaker or platform shaker:** A platform fitted with clips for grasping flasks (usually Erlenmeyer flasks), or with surfaces suitable for attaching flasks, with set or variable speed control. It is desirable to adjust the **shaking** speed for gentle, even agitation of suspension cultures.

**shoot:** Part of the plant axis which is generally above the ground, ascending, including branches, leaves and other organs. The shoot apex is an apical meristem, from which leaves and new shoots are formed.

**shoot apex or shoot tip:** The meristematic dome together with leaf primordia, emerging leaves and subjacent stem tissue. **Shoot tips** may be of apical (apical shoot tip) or lateral (lateral shoot tip) origin and are common explants in plant tissue culture (shoot tip culture). The term shoot tip should not be confused with the term meristem tip, which is much smaller and consists only of the meristematic dome and usually one pair of leaf primordia.

**shoot differentiation:** Generally refers to the development of growing points, leaf primordia and finally shoots from a shoot tip, axil or a callus surface in tissue culture.

Analogous to the development of apical or lateral shoots from a shoot tip or axil in vivo.

**shoot:root ratio:** The proportional amount of shoot and root growth. This ratio is determined by nutrients, especially hormones and environmental factors.

**shoot tip graft or micrograft:** The grafting of a very small shoot tip or meristem tip onto a prepared seedling or micropropagated rootstock in culture. Meristem tip grafting is used for in vitro virus elimination with *Citrus* and for other plants as an alternative to grafting in the greenhouse.

**short chain peptides:** Short peptide chains consisting of two to several amino acids, usually without enzymatic function.

**SI units:** The abbreviation for **Système International d'Unités**, a coherent and preferred system of units used for scientific purposes and currently replacing all others. The SI system consists of seven base and two supplementary units including the metre (m), kilogram (kg), ampere (a), second (s), mole (mol), kelvin (K), candela (cd), radian (rad) and steradian (sr). Derived units, many of which have special names, are used for other physical quantities. All have agreed symbols. Standard prefixes and abbreviations are used for decimal multiples of these units.

**sieve:** A utensil or process (**sieving**) to separate finer particles from coarser ones or solids from liquids.

**seive element:** See **phloem**.

**single cell line or cell strain:** A culture started from a single plant cell, usually from suspension cultures of single cells and small aggregates plated on solidified medium. The latter may incorporate a selective agent, from which tolerant or resistant individual cell lines or cell clones can be selected.

**sink:** A region receiving translocated sugars and other substances synthesized or stored elsewhere in the plant.

**Skoog, F. and C. Tsui (1948):** Established that shoot and root formation were chemically (hormonally) regulated in tobacco callus cultures.

**Skoog, F. and C.O. Miller (1957):** Established that the ratio of auxin to cytokinin was instrumental in controlling the formation of roots and shoots from tobacco callus and that this ratio constituted the basic regulatory mechanism involved in morphogenesis. They determined that a high auxin:cytokinin ratio promoted root formation, a low ratio promoted shoot development and intermediate ratios induced callus proliferation.

**slow-cooling:** Incremental temperature decrease used in cryopreservation of plant cells.

**soda glass or soda-lime glass:** Inexpensive, lower quality glassware alternative to Pyrex or borosilicate glass. It has less impact, abrasion and heat resistance and has a limited lifetime (ca. 1 year) for tissue culture purposes, unless coated periodically; as with a dimethyl dichloro-silane product.

**sodium (Na aw 22.98977 an 11):** A soft, white, reactive, metallic alkali group metal. It is a microelement necessary for some plants and beneficial to the growth of others. It is included as a salt in some plant tissue culture media.

**sodium chloride or table salt (NaCl mw 58.45):** A micronutrient **sodium** and **chloride** salt used in some plant tissue culture media. Currently there is much interest in the in vitro screening for salt tolerant cells, in the hopes of regenerating plants tolerant to high salinity soils, as much irrigated land globally is affected by sodium toxicity.

**sodium EDTA:** See **ethylenediaminetetraacetic acid, disodium salt**.

**sodium hydroxide (NaOH mw 40.01) or caustic soda:** A strong base used to raise the pH of media and to dissolve auxins and gibberellins in plant tissue culture work.

**sodium hypochlorite (NaOCl mw 74.44):** A frequently used plant tissue sterilant at 0.5–2% w/v. The usual source is dilute (10–21% v/v) commercial laundry bleach, sometimes with a surfactant or antioxidant added. Use of agitation or vacuum during surface sterilization is common. Tissue damage may occur if contact with tissue is prolonged. Thorough washing with sterile water generally follows treatment.

**soil free:** Both tissue culture and hydroponics utilize soilless nutrient solutions for plant growth in contrast to traditional soil use.

**solid or solidified medium:** A medium solidified with agar, a synthetic starch polymer or some other gelling agent. **Solid media** are widely used in plant tissue culture. For suspension cultures and for many research purposes liquid media are preferred.

**soluble salts:** **1.** Salts that dissolve in solution (nutrient medium). **2.** The nutrient solution (or water) ion concentration, measured as electrical conductivity.

**solute:** A substance that is dissolved in a fluid (solvent).

**solute potential:** See **osmotic potential.**

**solution:** **1.** The act of mixing, or the mixture formed when one substance (**solute**) is homogeneously mixed with another, usually liquid substance (solvent). **2.** The process of solving or the answer to a problem.

**solvent:** A substance which can dissolve other substances.

**somaclone:** A plant regenerated from a tissue culture originating from somatic tissue.

**somatic:** A plant body cell other than a germ cell. A non-sexual (vegetative) portion or process.

**somatic cell embryogenesis:** The production of embryos from somatic cells of explants (direct embryogenisis) or by induction on callus formed by explants (indirect embryogenesis). These two processes may not be materially different in results.

**somatic cell variant:** A somatic cell with unique features not shared by the others; as selected for in a screening trial that may follow a mutational event.

**somatic embryo or embryoid:** An organized embryonic structure morphologically similar to a zygotic embryo but initiated from somatic (non-zygotic) cells. These develop into plantlets in vitro through developmental processes that are similar to those of zygotic embryos.

**somatic hybrid:** A cell or plant product of somatic cell fusion; as the result of cell or protoplast fusion and implying genomic integration. The process is **somatic hybridization**.

**somatic mutation:** Mutation occurring in vegetative cells or tissues.

**somatic organogenesis:** The production of shoots, roots or other organs on somatic tissues of explants (direct organogenesis) or by induction on callus formed by explants (indirect organogenesis).

**sorbitol ($C_6H_{14}O_6$ mw 182.17):** A sugar alcohol which is the main translocatable carbohydrate in some plants. Occasionally it is added to plant tissue culture media.

**source:** Indicates a unique identity (by name or number) distinct from other **sources** with different origins within a cultivar or clone (source clone).

**source clone:** A source that originates from a single plant or explant within a clone; as from a specific virus tested (SVT) or specific pathogen tested (SPT) individual explant or plant.

**source plant:** A mother plant or donor plant from which an explant used to initiate a culture is taken.

**spatula:** A knife-like, sometimes spoon-like implement for scooping or spreading of chemical substances. It is commonly used to transfer small quantities of chemicals for weighing in media preparation.

**specialized:** Cells or organisms adapted anatomically or physiologically for particular functions or habitats. The act, process or adjustment is **specialization**.

**specific conductivity:** The reciprocal of a solution's resistance. This is measured in micro- or milli-ohm per cm or m at 25°C.

**specific pathogen tested (SPT):** A culture or plant that is apparently healthy and free from infection during propagation and maintenance. Refers to freedom from the specific pathogens for which a plant is being tested (such as bacteria or virus) and is not necessarily a guarantee of good health.

**specific virus tested (SVT):** A culture or plant that is apparently healthy and free from viral infection during propagation and maintenance. Refers to freedom only from the specific viruses for which the plant was tested and does not necessarily guarantee the absence of virus. See **source clone**.

**specificity:** A unique and obligatory relation between two or more processes, forms, etc.; as a causitive relationship.

**spent medium:** Medium discarded when a culture is subcultured. The implication is that the medium has been depleted of nutrients, dehydrated or accumulated toxic metabolic products.

**S phase:** The cell cycle phase during which DNA synthesis occurs.

**spike culture:** The aseptic culture of a grain inflorescence for microspore development.

**spindle:** A bundle of delicate microfibrils, extending between the two poles or asters in meiosis and mitosis, taking part in chromatid distribution.

**spirit lamp, alcohol lamp or burner:** A glass or metal container, with a wick emersed in ethanol or some other spirit (alcohol) and protruding at one end through a cork or other sealant. This device is used to produce a flame; as for sterilizing tools used in tissue culture.

**spongy parenchyma:** The parenchyma cells of a leaf bearing chloroplasts. These are variably shaped and lobed and often loosely packed with many air spaces. They are located below the palisade parenchyma and together they make up the mesophyll.

**spontaneous fusion:** Uninduced protoplast fusion which may occur between freshly isolated protoplasts or following adhesion of adjacent cells during enzymatic cell wall degradation.

**spontaneous variation:** The variation in plant populations derived from tissue cultures not exposed to mutagens but occurring as a result of the culture conditions.

**sporophyte:** The diploid (2n) generation of the life cycle of plants, with alternation of generations, during which spores are produced.

**sport:** An individual or portion thereof distinguished by a spontaneous mutation. **Sports** are sometimes of great horticultural worth. Alternatively, they may be disadvantageous and may be rogued during agronomic production.

**spray:** **1.** A group of flowers or leaves and small branches (branchlets). **2.** A fine mist.

**sprig:** A small branch, shoot or twig.

**sprout:** The first shoot to grow from a seed or root or in the process of germination and development.

**SPT:** See **specific pathogen tested**.

**SSEM:** See **serologically specific electron microscopy**.

**stable:** Durable, permanent or resistant to change; as some cell lines in vitro are genetically more stable than others.

**stages of culture (I-IV):** **Stage I:** Aseptic explantation or establishment of the explant in culture. **Stage II:** Multiplication of the propagules. **Stage III:** Rooting of the propagules and preparation for transplant to soil. **Stage IV:** Establishment of Stage II or III propagules ex vitro in soil or potting mix.

**stalk:** A plant stem or support structure (filament, pedicel).

**stamen:** The floral organ producing microspores (pollen) in angiosperms. It is composed of a stalk (filament) at the top of which are two pollen sacs (anthers) containing pollen. Collectively the **stamens** form an androecium.

**starch (($C_5H_{10}O_5$)n mw (102.09)n):** A complex, insoluble polysaccharide of glucose units (amylose, amylopectin; glucose branching arrangements differ) used as a carbohydrate storage substance by plants and deposited in the form of grains. The latter are formed in chloroplasts as a product of photosynthesis or in colorless plastids (leucoplasts) in storage tissue (cotyledons, endosperm, roots or other). Starch stains blue-black with an iodine solution (a starch indicator). This indicator can be made up using 1 g I and 2 g KI in 300 ml water. This solution should be stored in the dark.

**stationary culture:** A non-agitated culture. The antonym is shake culture.

**stationary phase:** Batch suspension cultures enter this phase, of little new growth, when a factor in the medium becomes limiting. Subculture is generally indicated at this point.

**statolith:** A starch grain or other solid plant cell inclusion located on the lower side in certain root cap cells (statocysts). These are moved by gravity and believed to provide the stimulus for plant geotropic responses. The roots of ex vitro transplants sometimes exhibit negative geotropism. This response may be related to abnormal statolith formation.

**stele:** The central cylinder or core within the cortex, surrounded by endodermis, in the stems and roots of vascular plants. Stele structure differs in different plant groups, consisting of vascular tissue, pericycle and in some steles, pith and pith rays.

**stem:** **1.** The main shaft (stock) of a plant bearing buds, leaves, fruits and other parts. Most are ascending and aerial. Some are subterranean (rhizomes), and are distinguishable from roots by buds, leaves (may be reduced to scales) and stem-type vasculature. **2.** A stalk or support structure; as for leaves and flowers.

**stem cutting:** A section of stem or shoot divided for rooting purposes. A commonly employed propagation procedure.

**sterile:** **1.** Unable to bear fruits, crops or offspring. **2.** Free from infectious matter or agents. A **sterilizer** or **sterilant** makes things sterile.

**sterile transfer:** The process of aseptically explanting or subculturing; moving plant material from one location or culture to another under aseptic conditions.

**sterilization:** The process of making things **sterile** through: **1.** Rendering plants non-reproductive. **2.** Killing or excluding microorganisms or their spores with heat, filters, chemicals or other **sterilants**. Dry heat sterilization is useful for metal instruments and glassware. Foil-wrapped items are subjected to 150°C for a minimum of 3 hours in a hot air oven. Steam sterilization or autoclaving is useful for nutrient media, distilled water, paper products and glassware. Solutions contained in glass flasks, plugged with cotton and capped with foil are subjected to 1.05 kg/cm$^2$ (121°C) for 10 to 20 minutes (depending on the total volume and how it is distributed). Foil wrapped paper products or foil capped glassware are autoclaved in the same way. Filter sterilization through a "Millipore-type" membrane is useful for thermolabile solutions such as those containing vitamins and urea. Chemical sterilization (most commonly sodium hypochlorite) is useful for plant materials in preparation for excision (surface sterilization) and for working surfaces. Some medium constituents that are thermolabile or insoluble in water may be sterilized through dissolving in organic solvents, such as chloroform or alcohol. They are then dispensed to sterile filter paper for solvent evaporation and the filter paper with the residue is added to or below the sterile medium. Metal instruments are flame sterilized by immersion in 70% ethanol until required, then flaming.

**steroid:** A saturated hydrocarbon (17 carbons arranged in rings) which is derived from cholesterol, of diverse biological activity. Many are hormones, some are toxins or poisons which have medicinal properties. The objective of much tissue culture effort has been to induce the synthesis in culture of large quantities of select **steroids**.

**Steward, F.C., M.O. Mapes, and K. Mears (1958):** Were among the first to (independently) observe adventive embryogenesis in cell cultures (of carrot).

**stigma:** The expanded style apex and pollen-receptive part of the pistil on which pollen grains germinate.

**stimulus:** An environmental or internal change triggering or inducing changes in the activities of organisms or portions thereof.

**stock:** **1.** The root and a portion of the stem (rootstock) of a plant to which is grafted a part of the same or another plant (scion). **2.** A group of closely related plants.

**stock plant:** The source plant from which cuttings or explants are made. These are usually maintained carefully in an optimum state for (sometimes prolonged) explant use. Preferably they are certified, pathogen-free plants.

**stock solution:** A solution, usually concentrated (10 to 100 times the final medium concentration), of select medium constituents that are grouped for compatability to avoid precipitation and prepared before hand to save time during medium preparation. Usually they are frozen or stored in the refrigerator and portions are utilized as media are prepared.

**stolon:** A slender, horizontal stem rooting at the nodes; as does a strawberry runner or offset from a leek.

**stoma, pl. stomata or stomates:** An opening or pore between two specialized epidermal cells (guard cells) whose movements open and close the pore, based on changes in turgidity. Epidermal cells adjacent to the guard cells may differ in size or arrangement from the rest of the epidermal cells; these cells are called subsidiary cells. **Stomata** are located on leaves and young stems and in lesser numbers on other above ground structures in green plants. They penetrate the epidermis and allow gas exchange between the air and inner tissues necessary for photosynthesis and transpiration. In vitro shoots and plantlets generally have stomata that are fixed open and incapable of closing.

**stomatal complex:** Includes the stoma together with the guard cells and, when present, the subsidiary cells.

**stomatal index:** The number of stomata divided by the number of stomata plus the number of epidermal cells, all multiplied by 100, calculated per unit area (usually per $mm^2$). This value has been found to be reasonably constant for a particular species and is useful in comparing leaves of different sizes. During leaf development this value is affected by relative humidity and perhaps by light intensity.

**streptomycin ($C_{21}H_{39}N_7O_{12}$ mw 581.58):** An antibiotic obtained from *Streptomyces griseus*. This antibiotic is sometimes included in plant tissue culture media to control contamination when other means may not be appropriate.

**stroma: 1.** The chloroplast matrix in which the grana are embedded containing enzymes and photosynthetic reagents. **2.** A mass of fungal hyphae which may give rise to fruiting structures.

**style:** The elongated portion of the pistil, connecting the carpel to the stigma in a flower, down which the pollen tubes grow after pollination.

**subdivide:** To divide into several parts; as in **subdivision** and subculture to fresh nutrient medium.

**subculture or passage:** A culture derived from another culture or the aseptic division and transfer of a culture or a portion of that culture (inoculum) to fresh nutrient medium. Subculturing is usually done at set time intervals, the length of which is called the subculture interval or passage time.

**suberin:** A phenolic condensation product containing fatty acids which is a component of endodermal Casparian strips, and is also present in cork cell walls, rendering them impervious to water. The deposition process is **suberization.**

**subline:** A cell line regenerated from a unique cell line of a hybrid callus colony.

**$N^6$-substituted:** Describes a purine in which a proton of the $N^6$ amine group is replaced by another group or side chain. Examples are 6-benzylaminopurine and 6-furfurylaminopurine.

**substrate: 1.** The surface that something is on. Synonymous with substratum. **2.** The material that forms the growth medium for a microorganism or plant tissue culture, or upon which it occurs. **3.** The substance upon which an enzyme acts.

**succinic acid ($C_4H_6O_4$ mw 118.09):** A crystalline, white dibasic acid which is one of the TCA cycle intermediates. It is employed in the production of dyes, lacquers and other products and is occasionally included in plant tissue culture media.

**sucker:** A shoot from the stem base region or the roots of a plant.

**sucrose (cane or beet sugar, $C_{12}H_{22}O_{11}$ mw 342.30):** A disaccharide of glucose and fructose found in plants. It is the major transport sugar in plants. It is the most commonly employed carbohydrate source and osmoticum in plant tissue culture media and is used at 20–40 g/liter.

**sugar:** Any sweet, soluble, crystalline, lower molecular weight carbohydrate; the monosaccharides, oligosaccharides and their derivatives, especially sucrose.

**sugar alcohol:** A monosaccharide with an aldehyde group reduced to an alcohol including sorbitol, glycerol and inositol. Inositol is usually included (ca. 100 mg/liter) and sorbitol is less often used in plant tissue culture media.

**sulphur or sulfur (S aw 32.064 an 16):** A nonmetallic element. As a macroelement it is a constituent of some amino acids and takes part in many vital plant reactions. It is a necessary addendum to nutrient media used for plant tissue culture and is supplied as the **sulphate** compounds **manganese sulphate**, **magnesium sulphate**, **zinc sulphate**, **copper sulphate**, **iron sulphate** and others.

**Sunderland (1977–1979):** See **serial float culture**.

**supraoptimum:** An amount (level) greater than required; as an inhibitory concentration of an exogeneous growth factor.

**surface sterilization:** The removal of plant surface microflora prior to aseptic excision of explants. Surface sterilization is accomplished by immersion of tissue in one of many sterilants, such as calcium hypochlorite or sodium hypochlorite, hydrogen peroxide, mercuric chloride, silver nitrate or bromine water for an empirically determined period of time. Sometimes a preceeding ethanol dip or spray is useful, as is the addition of a surfactant to the sterilant. Agitation or vacuum may improve the efficiency of sterilization. Thorough washing in sterile distilled water usually follows, to remove the sterilant, although in some cases this is not necessary (hydrogen peroxide).

**surface tension:** Tension exerted by a liquid surface due to molecular cohesion and apparent at liquid boundaries.

**surfactant:** A surface active agent or wetting agent; as is Tween 20 or Tween 80, Teepol, Lissapol F, Alconox, etc. These agents act by lowering the surface tension so are common addenda to solutions used to surface sterilize plant materials prior to aseptic excision of explants.

**suspension culture:** Cells and groups of cells (aggregates) dispersed in an aerated, usually agitated, liquid culture medium. These are obtained by adding friable callus to the medium. The plant species, explant type and treatment; the composition of the nutrient medium; and many other features determine the size and nature of the cell aggregates, which may appear different during each culture growth phase. These cultures are used to study cell division, differentiation and metabolism, in secondary product synthesis or form the basis for single cell lines, callus cultures, somatic cell embryogenesis and many other culture purposes.

**SVT:** See **specific virus tested**.

**synchronized cells:** Synchronized mitosis in a group of cells in culture by natural or artificial means.

**synchronous culture:** A plant cell or microbial culture treated in such a way as to have all (or most) cells or individuals in the same stage of development or mitosis. This can be achieved in various ways including via temperature variation and nutrient limitation.

This results in a stepwise increase in cell numbers and provides the opportunity to examine changes (biochemical or other) on a large sample size.

**synergidae or synergids:** At the micropylar end of an embryo sac, these two haploid nucei are located close to the egg nucleus, and with it they form the egg-apparatus of the angiosperm female gametophyte. Synergidae function is unknown and they abort soon after fertilization.

**synergism:** The combined effect of two or more agents, such as medium constituents, which together have a greater effect than the sum of the separate effects.

**syngraft:** See **isograft**.

**synkaryon or synkaryocyte:** Refers to a heterokaryon in which the nuclei have fused.

**synthesis:** Compound formation or their construction from elements or simpler compounds; as in synthesis of plant proteins from amino acids.

**system:** **1.** The method, technique or arrangement; as in a closed culture system. **2.** An organizational or functional unit; as of the metric or vascular systems.

**Système International d'Unités:** See **SI units**.

**systemic:** Internal, or spread; as via translocation of a pathogen, throughout the entire organism.

# T

**2,4,5-T:** See **(2,4,5-trichlorophenoxyacetic acid)**.

**tagged:** Labelled or otherwise identified, enabling recognition or tracing during an investigation.

**tapetum:** The innermost wall layer in an anther; absorbed as food as the pollen grains mature.

**taproot:** A stout, vertical, main root from which smaller lateral branches emerge. It may become enlarged, containing stored food materials; as in the edible portion of a carrot.

**tare or tare weight:** The weight of the weighing paper or other container that is deducted from the gross weight of the substance being weighed mechanically, using the balance, or arithmetically.

**TCA cycle:** See **tricarboxcylic acid cycle**.

**telophase:** The last meiotic or mitotic phase during which two nuclei form and cytokinesis takes place.

**tent:** The enclosure or process of including plants, such as ex vitro plantlets, under plastic or glass to maintain elevated relative humidity.

**teratogenic:** Refers to an agent which induces or increases the incidence of gross structural abnormalities (**teratomata**) in an individual or a population; as do x-rays.

**terpene:** An unsaturated hydrocarbon unit that is a constituent of plant oils, resins and various secondary products.

**terpenoid:** Any of a large and diverse class of compounds derived from multiples of isoprene units. In this class are the terpenes, a group of hydrocarbons present in many fragrant essential oils of plants; as pinine (pine oil), found in terpentine, limonene (lemon oil), found in citrus fruits and menthol (mint oil) found in mint.

**terminal:** **1.** Located at the apex, tip or end. **2.** The end of a process such as development.

**test tube:** A thin glass tube open at one end and used with an appropriate closure for plant tissue culture and other, usually scientific, purposes.

**test tube fertilization:** Pollination, followed by fertilization, occurring in vitro.

**testa:** The protective seed coat, formed from the integuments of the ovule.

**tetrad:** A group of four haploid cells formed by meiosis; as the four microspores formed in pollen production.

**tetramethylrhodamine isothiocyanate:** See **rhodamine isothiocyanate**.

**tetraploid:** A chromosome number four times the haploid number (4n) (polyploid).

**tetrazolium or triphenyl (2,4,5)tetrazolium chloride (TCC, $C_{19}H_{15}NCl$ mw 334.81):** Used in a quantitative assay of cell viability. It becomes reduced to a red, water-insoluble form, formazan, in the presence of oxidative metabolism. It is then dissolved in alcohol and measured spectrophotometrically.

**thermic shock:** Exposed to a reduced temperature for several days, an allegedly beneficial pretreatment for immature flower buds or anthers prior to anther culture or pollen culture.

**thermolabile or heat labile:** Destroyed or altered during heating; as some viruses are killed during heat therapy (thermotherapy) of infected plants, or as some medium components which are destroyed by autoclaving and must be filter sterilized.

**thermostat:** A device which maintains temperature within fixed limits.

**thermotherapy or heat therapy:** Exposure of plants to elevated temperatures as a treatment for virus or mycoplasma elimination. Thermotherapy is used alone or in conjunction with meristem culture or meristem tip culture for virus elimination.

**thiamine or aneurine ($C_{12}H_{17}ON_4SCl$ mw 300.82):** Vitamin $B_1$, a coenzyme involved in carbohydrate metabolism. A common constituent of nutrient media used for plant tissue culture. It is usually added as **thiamine hydrochloride**.

**thiamine hydrochloride or aneurine hydrochloride ($C_{12}H_{17}ON_4SCl.HCl$ mw 337.28):** A thiamine (vitamin $B_1$) salt commonly included in plant tissue culture media.

**thiol compound:** An organic substance possessing an -SH (sulfhydryl) group capable of forming -S- esters.

**2-thiouracil:** An anti-viral agent sometimes included in initiation medium for virus elimination from infected explants. Of uncertain value, its mode of action is unclear, but may involve inhibition of viral RNA synthesis.

**thiourea ($CH_4N_2S$ mw 76.12):** An occasional additive in plant tissue culture media as a source of reduced nitrogen.

**thrip:** An insect of the order Thysanoptera, common greenhouse and garden pests. **Thrips** may infest plant tissue culture facilities and culture containers, spreading fungal and bacterial contaminants. Insecticidal paint or shelf paper is useful in their control.

**thylakoid:** In photosynthetic organisms these vesicle membranes, which make up the grana, bear the photosynthetic pigments.

**Ti plasmid:** A portion of the genome of the bacterium *Agrobacterium tumefaciens*, the agent involved in crown gall disease. This plasmid is a useful (experimental) vector for the transfer of genetic information into plant cells.

**TIBA:** See **2,3,5-triiodobenzoate**.

**tin (Sn aw 118.69 an 50):** A soft, silvery white, malleable and ductile metal. It is sometimes added to plant tissue culture media as **tin (stannous) chloride**.

**tissue:** A group of cells with similar origin and structurally organized into a functional unit. The organs of multicellular organisms are made up of combinations of tissues, of one or more types of cells.

**tissue culture:** **1.** A general term for the cultivation of plant parts (cells, tissues or organs) under aseptic conditions in synthetic medium in vitro. It also refers to the cultures themselves. **2.** A technique for vegetative propagation of plants.

**tissue explant:** An excised plant portion of tissue used to initiate a culture.

**tobacco (*Nicotiana*) species:** One of a limited number of crop species extensively grown in culture because of its ease of manipulation and readiness of regeneration. MS (1962) medium was formulated in experiments with tobacco callus.

**$\alpha$-tocopherol ($C_{29}H_{50}O_2$ mw 430.69):** Vitamin E, an occasional additive in plant tissue culture media.

**$\alpha$-tocopheryl acetate ($C_{31}H_{52}O_3$ mw 472.73):** A form of vitamin E occasionally used in plant tissue culture media.

**tolerance:** The capacity to endure or adapt; as to specific environmental changes, selective agents or to culture conditions.

**tonoplast:** The specialized membrane separating the central plant cell vacuole from the rest of the cytoplasm.

**topophysis:** See **determination**.

**totipotency:** The potential (**totipotential**) or inherent capacity of a plant cell or tissue to develop into (recreate) an entire plant if suitably stimulated. Totipotency implies that all the information necessary for growth and reproduction of the organism is contained in the cell. Although theoretically all plant cells are **totipotent** the meristematic cells are best able to express it.

**toxic:** Poisonous; as are some chemicals (**toxicants**) or any substances present to excess and detrimental to normal plant function or growth.

**trace element:** See **microelement**.

**tracer:** A substance that can be followed within a reaction or an organism; as radioactive isotopes and certain dyes.

**tracheary element:** A water-conducting, lignified xylem cell, as tracheid or vessel element. The secondary wall thickenings of **tracheary elements** are characteristic; annular, spiral, reticulate or scalariform.

**tracheid:** In vascular plants, a non-living xylem element with thick, lignified, pitted walls and tapered ends. It functions in mechanical support and in water conduction. Tracheids are considered to be more primitive than vessels because they are longer, smaller in diameter and lack perforation plates.

**trait:** A characteristic feature.

**transcription or genetic transcription:** DNA dependent RNA synthesis. The process by which cellular RNA molecules are synthesized as determined by homologous DNA sections where their sequence recurs in complementary form (the template or transcription strand).

**transfer:** **1.** Culture initiation; the first placement of an explant in culture. **2.** Subculture; relocation of cultures to fresh nutrient medium. **3.** Chamber; laminar air flow cabinet or hood (transfer hood) or room (transfer room) in which this is accomplished. **4.** The act or process of **transferring**.

**transfer RNA:** See **ribonucleic acid, transfer**.

**transfer room:** A small room, sometimes sterilized internally by a bactericidal lamp (u.v. irradiation) and provided with clean (filtered) air; employed for sterile transfer work.

**transitional:** **1.** A term applied to organs or plantlets undergoing phenotypic change; as during acclimatization after ex vitro transplant. Persistent organs (from culture) do not change substantially; new organs produced after transplant range from culture type to control type in anatomy and physiological performance, and so are termed the transitional organs. **2.** An area between the stem and root of plants of intermediate tissue arrangement, termed the transitional region.

**translocation:** **1.** The movement (transport) of dissolved substances, within plants via the conducting tissues xylem and phloem (in vascular plants). The xylem conducts water and mineral salts from the roots in the transpiration stream to the aerial parts of the plant. The phloem conducts mainly sugars or other carbohydrates inside a plant from one region (usually the leaves) to growth or storage areas. A narrower use may imply phloem transport only. **2.** An interchange of chromosomal segments between non-homologous chromosomes.

**transmission electron microscope (TEM):** A microscope that uses an electron beam to image objects and so has a much greater resolving power than a light microscope. The electron beam is passed through the object, which must be very thin, and scattered in a characteristic way. The scattered electrons are focused through a magnetic or electrostatic system and imaged on a florescent screen.

**transpiration:** The movement of water vapor through evaporation from a plant leaf to the atmosphere. Transpiration occurs mainly through leaf stomata but also through cuticles. Water movement is down water potential gradients from high potentials in soil and roots to low potentials in leaves and the atmosphere. Wilting (loss of turgor) can occur if the transpiration rate exceeds water uptake by the roots.

**transplant:** **1.** To relocate or remove to a new growing place. **2.** The cultured tissue or explant, relocated or transferred to a new site (in vitro). **3.** Stage IV; the transfer of plantlets or shoots ex vitro from aseptic culture to soil. The process is **transplantation**.

**transplant shock:** Refers to the stress involved when Stage II or Stage III cultures are transplanted to soil (Stage IV); many or all of the regenerated shoots or plantlets may die if suitable care is not taken to acclimatize them gradually to the soil environment, particularly prevention of water stress.

**tricarboxylic acid cycle (TCA cycle) or citric acid cycle or Krebs cycle:** In aerobic organisms this amphibolic cyclic sequence of reactions is almost universal and is present in the mitochondria; energy present in glucose is converted to ATP. During the cycle the acetyl portion of acetyl CoA is oxidized to hydrogen ions and carbon dioxide. NAD and flavoprotein are concomitantly reduced, eventually prviding energy for the formation of ATP via the respiratory chain and oxidative phosphorylation. The overall process of glucose oxidation results in the production of 38 ATP and 6 $CO_2$ from each sugar molecule.

**(2,4,5-trichlorophenoxy)acetic acid (2,4,5-T, $Cl_3C_8O_3H_5$ mw 255.49):** A synthetic hormone analog of the auxin type. It is used as a weedkiller and defoliant. It is sometimes used in plant tissue culture media. It dissolves in base (ca. 1M KOH or NaOH).

**trichome:** A specialized epidermal or subepidermal outgrowth, as are hair, scales, etc., with varied form and function. These are usually supplied with a vascular system. See **hair**.

**trifoliate:** A branch with three leaves, or leaves in groups of three.

**trifoliolate:** A leaf composed of three leaflets.

**trihydroxy alcohol:** An alcohol possessing three -OH groups. See **glycerol**.

**2,3,5-triiodobenzoate (TIBA mw 499.81):** An inhibitor of auxin movement or transport (antiauxin), sometimes included in plant tissue culture media for its growth promoting effects.

**triphenyl (2,4,5-) tetrazolium chloride (TTC, $C_{19}H_{15}NCl$ mw 334.81):** Used in a quantitative method to assay cell viability based on reduction of TTC, by oxidative metabolism, to a water insoluble red compound (formazan). Formazan is then dissolved in ethanol and quantified spectrophotometrically.

**triplet:** A sequence of three nitrogenous bases in a nucleic acid chain. See **codon**.

**tRNA:** See **ribonucleic acid, transfer**.

**true to type:** Applied to a plant or propagation source this term denotes correct cultivar identification and lack of variation in productivity or performance. Verification is determined visually by an expert or through biochemical, serological or other means.

**tryptamine or 3-(2-aminoethyl)indole ($C_{10}H_{13}N_2$ mw 161.23):** An occasional addition to plant tissue culture media as a source of reduced nitrogen.

**tryptophan (Trp, $C_{11}H_{12}O_2N_2$ mw 204.23):** An amino acid precursor of indole compounds such as IAA. Occasionally added to plant tissue culture media.

**trypan blue ($C_{34}H_{24}N_6Na_4O_{14}S_4$ mw 960.83):** A dye used in a quantitative method to assay cell viability. The method is based on the ability of live protoplasts to exclude the dye while dead protoplasts cannot.

**tuber:** A fleshy and enlarged underground stem or root for food storage, overwintering and vegetative reproduction. Stem tubers form at the tip of a thin stolon and have minute buds or eyes; as do potatoes. Root tubers are adventitious roots with no buds; as in *Dahlia* spp.

**tumble tube:** A glass tube closed at both ends with a side-neck opening which is commonly attached to a slowly revolving platform. Suspension cultures are agitated and aerated as medium flows from one end of the tube to the other as it is inverted. These tubes and a rotating platform were designed by F.C. Steward et al. (1952).

**tumor or tumour:** A swelling or growth of new (neoplasmic) tissues that may be anatomically or physiologically abnormal.

**tumor inducing principle (TIP):** The plasmid carried by *Agrobacterium tumefaciens*, the crown gall organism. Through incorporation into the host genome the host tissue is transformed into tumor tissue.

**tunica:** One of two peripheral, anticlinally dividing cell layer(s), (designated LI and LII) forming a mantle over the interior (corpus) in the shoot apex.

**Tunica-corpus theory:** The theory of Schmidt (1924) which suggests that the exterior plant is formed from the tunica, the corpus gives rise to the internal plant body.

**turbidity:** The degree of cloudiness present in a liquid, which may be indicative of the amount of cell growth or of medium contamination.

**turbidostat:** An open continuous culture system wherein the inflow of fresh medium is controlled by the **turbidity** of the culture, a function of the amount of cell growth. Balancing the fresh medium inflow is a regulated outflow of cells and spent medium, restoring the original turbidity level.

**turgid:** The rigid condition of a cell resulting from internal water pressure (turgor pressure).

**turgor:** The state of cell rigidity due to internal water pressure. An essential feature in young cell expansion, stomate opening, phloem translocation and in mechanical support of succulent plant parts.

**turgor pressure:** The pressure of cellular contents exerted against the plant cell wall. The counterpressure exerted by the cell wall on the protoplast is wall pressure.

**Tween 20 (polyoxyethylene sorbitan monolaurate mw 1227.54):** The brand name for a wetting agent or surfactant which breaks the surface tension of tissues. Commonly added to disinfecting solutions to make them more effective.

**twig:** A shoot or small branch of a tree.

**tyrosine (Tyr, $C_9H_{11}NO_3$ mw 181.19):** An amino acid, sometimes used in plant tissue culture media as a source of nitrogen.

# U

**ubiquitous:** Occurs everywhere, as do bacteria in the environment.

**ultrasonic cleaner:** A device to induce high frequency vibration of materials, removing adhering substances from surfaces by mechanical action. This device is useful for cleaning glassware and for disinfecting plant material.

**ultraviolet light (u.v.):** Radiation with wavelength (100–400 nm) at the violet end of the visible spectrum. Generated by mercury vapor lamps, it is sometimes used in tissue culture for its mutagenic properties or to reduce ambient contaminants in work areas due to its bactericidal properties. Exposure is harmful to the eyes and skin and is to be avoided by workers in plant tissue culture facilities.

**undefined:** A medium or substance added to medium in which not all of the constituents or their concentrations are chemically defined; as media containing coconut milk, malt extract, casein hydrolysate, fish emulsion or other complex addenda. The aversion to undefined materials or media is based on a lack of control implied by their use and a concern with their variability from batch to batch.

**undifferentiated:** Lacking in specific functional role or anatomical structure; as are meristematic cells.

**unorganized growth:** In vitro formation of tissues with few differentiated cell types and lacking recognizable structure; as with many calli.

**unstable:** Changeable, uncertain; as chromosome **instability** is a recurrent problem with somatic hybrids.

**urea ($CH_4N_2O$ mw 60.06):** A white crystalline nitrogenous solid that occurs in urine. It is used as a fertilizer and is also added to some plant tissue culture media as a source of nitrogen.

**uronic acid:** A member of the most important group of sugar acids, the common constituents of polysaccharides. In uronic acids the carbon atom carrying the primary hydroxyl group (the $CH_2OH$ end) is oxidized to a carboxyl group (COOH); as glucuronic and galacturonic acids.

**u.v.:** See **ultraviolet light**.

# V

**vacuole:** A cavity in the cell cytoplasm bounded by a membrane, the tonoplast, and containing watery fluid or cell sap. In this cell sap are storage substances, dissolved gases, pigments and waste substances. Many mature plant cells have a single vacuole that occupies most of the cell volume. **Vacuoles** control cell turgor via osmotic water exchange.

**vacuum:** **1.** A space partially exhausted of air. **2.** The creation of space empty of air (or matter). Commonly employed in preparing plant material for culture to enhance disinfection, through use of a vacuum pump or aspirator pump.

**valence:** The capacity of atoms (or groups thereof) to combine by sharing electrons. Some elements have more than one valence state; these are interchangeable and depend on electron transfer.

**valine (Val, $C_5H_{11}NO_2$ mw 117.1:** An amino acid found in seeds and proteins. Occasionally added to plant tissue culture media.

**van Overbeek, J., M.E. Conklin and A.F. Blakeslee (1941):** First to demonstrate the growth-promoting effect of coconut milk (on excised *Datura* embryos).

**variation:** The difference in phenotype (anatomy or physiology) among members of a clone (**variants**) or of a species or a group (apart from developmental or ontogenetic changes). This may be the result of genetic (mutational), environmental or a combination of influences.

**variegated:** Plants having different colors; as both green and albino tissue. This variation may result from viral infection; nutritional deficiency; or be under genetic or physiological control.

**variety:** A subgroup within a species different from the rest of the species in some minor characteristic(s).

**vascular:** Pertaining to the conductive and strengthening tissues of plants, the xylem and phloem, or to possession of these tissues as in the vascular plants; angiosperms, gymnosperms, ferns, etc.

**vascular cambium:** The lateral meristem that divides to form secondary xylem to one side and secondary phloem to the other in plants with secondary thickenings.

**vector:** **1.** Any agent (biotic or abiotic) that carries (carrier) or contains (host) an infectious agent and transmits it from one organism to another, and so spreads a disease. **2.** A pollen vector carries pollen from one plant to another. **3.** The plasmid or nucleic acid vector used in the insertion of genes into cells in genetic engineering.

**vegetative:** Relating to nutrition, maintenance and growth (non-sexual reproduction), in contrast to reproductive activity. During the vegetative phase an organism is not occupied in reproductive activities.

**vegetative propagation (vegetative reproduction):** The propagation of plants by means other than by seeds. Asexual detachment of some specialized somatic portion of the plant body (bulb, tuber, leaf, etc.) followed by its development into an entire plant.

**vermiculite:** Any of many minerals (commonly altered micas) whose granules can expand greatly, becoming highly absorbent. Used alone or more commonly as a component of potting mix.

**vernalin:** An hypothetical hormone-like substance found in plant meristematic regions, produced by vernalization. This substance is apparently graft transmissible, but has not yet been identified. Different cold requiring species may form different substances during vernalization.

**vernalization:** **1.** Low temperature exposure required by some plants to induce bud break or flowering. **2.** Low temperature exposure of seeds to shorten the time required for flowering. In some species vernalization requirements may be replaced in whole or in part by auxin, gibberellin, kinetin, RNA or vitamin E. Note: The low temperature treatment of seeds to induce germination is called stratification.

**versene:** See **ethylenediaminetetraacetic acid**.

**viable:** Capable of germinating, living, growing or sufficiently developed physically as to be capable of living.

**viability test:** Assay of the the number or percent of living cells or plants in a population that has been given a specific treatment; as do tests for cell viability after cryopreservation treatment.

**victorin:** A host specific toxin produced by *Helminthosporium victoriae* utilizable for screening oat cells for resistance to the pathogen.

**virazole or ribavirin or 1-β-D-ribofuranosyl-1,2,4-triazole-3-carboxamide:** A purported anti-viral agent of uncertain value. It has been incorporated into initiation medium for virus elimination, usually in conjunction with thermotherapy and meristem or meristem tip culture.

**viroid:** An infectious agent causing plant diseases, smaller than viruses and composed only of RNA, with no coat protein.

**virus:** One of many submicroscopic (light microscope) self-replicating, (within host cells) infectious agents capable of passing through a bacterial filter. **Viruses** are composed of nucleic acid (DNA or RNA) encased in a protein sheath (capsid). Viruses require intact host cells for replication (obligate intracellular parasite). They may or may not cause obvious symptoms in infected plants. Some can be seen with the electron microscope or detected by a number of assays, including serological tests or by grafting to indicator plants.

**virus elimination:** Thermotherapy, chemotherapy and meristem or meristem tip culture, used alone or in combination have been employed for the elimination of systemic viruses from plants. Plants must repeatedly test negatively for the virus in question for assurance that it has been eliminated.

**virus-tested or virus-free:** A plant that appears healthy and repeatedly tests negatively for the presence of one or more identifiable viruses. Such a plant may then be used as a stock or donor plant (explant source) for propagation purposes, and may be certified as virus tested (certified virus tested). The term "virus-free" is incorrect in most cases, as such a plant may contain one or more viruses which have not been assayed.

**vital stain:** A dye solution of which the dye is taken up only by living cells and used for indicator purposes. Florescent dyes with specific molecular affinities are often used for the tagging of discrete cellular components (membranes, macromolecules, organelles, antigens, etc.).

**vitamin:** One of many trace organic substances some of which function as components of certain coenzymes or cofactors for vital metabolic functions and are required in the normal diet of species which are unable to synthesize them. Some are water soluble (vitamin B complex and ascorbic acid) and the others are fat soluble.

**vitamin B complex:** Includes a large group of water soluble vitamins that function as coenzymes; thiamine ($\mathbf{B_1}$), riboflavin or vitamin G ($\mathbf{B_2}$), niacin or nicotinic acid ($\mathbf{B_3}$), pantothenic acid ($\mathbf{B_5}$), pyridoxine ($\mathbf{B_6}$), cyanocobalamin ($\mathbf{B_{12}}$), biotin or vitamin H, folic acid or vitamin M (**Bc**), inositol, choline and others.

**vitamin $\mathbf{B_x}$:** See **para-amino-benzoic acid**.

**vitamin C:** See **ascorbic acid**.

**vitamin D:** A fat soluble vitamin composed of a group of related steroids. Present in all plants.

**vitamin E:** See **tocopherol**.

**vitamin G:** See **riboflavin**.

**vitamin H:** See **biotin**.

**vitamin K:** A fat-soluble quinone involved in photosynthetic electron transfer in plants and blood clotting in animals.

**vitamin PP:** See **niacin**.

**vitrified or water soaked:** Cultured tissue having leaves and sometimes stems with a glassy, transparent or wet and often swollen appearance. Eventually shoot tip, then leaf necrosis occurs. The process is **vitrification**. This is a general term for a variety of physiological disorders.

**volatile:** A liquid readily vaporizing or evaporating at relatively low temperature and pressure.

**volume:** The space occupied as measured in cubic units; mass or bulk.

**volumetric flask:** A glass vessel, precisely calibrated to contain a known volume, used for solution preparation and storage.

**v/v:** May indicate simple proportion; as 3:1 (v/v). May indicate percent volume in volume; as the number of $cm^3$ (mls) of constituent in 100 $cm^3$ (mls) solution.

# W

**W:** An abbreviation for the **White, P.R. (1954)** medium formulation.

**wall pressure:** The pressure exerted by cell walls against cellular contents. It is the inverse of turgor pressure.

**watch glass:** A small curved glass dish sometimes used as small scale culture containers. They are usually covered with a cover glass or enclosed in petri dishes. Their use has been largely supplanted by petri dishes, culture tubes, etc.

**water of hydration:** The amount of water chemically bound to a substance. This amount may be variable and must be taken into account when solutions of salts are prepared; as in medium preparation.

**water potential:** Governs the conduction of plant water from regions of high to low water potential (towards increasing solute concentration) and is a measure of the free energy status of water in a system. The total water potential of the system is the sum of the osmotic component (dependent on solute concentration), the matric component (dependent on water-binding substances and surface forces) and the pressure component (usually turgor pressure).

**Went, F.W. and K.V. Thimann (1937):** Discovered the auxin 1*H*-indole-3-acetic acid (IAA) in fungi and summarized the knowledge of this auxin up to that time.

**wetting agent or surfactant:** A substance that reduces the surface tension of a liquid so improves surface contact; as Tween 20, Triton X-10 (brand names) which added to disinfecting solutions promote the disinfection process.

**wet weight:** The gross weight or weight of fully hydrated tissue.

**White, P.R.:** Produced the first successful plant tissue cultures (organ cultures) in the early 1930's. He isolated and grew tomato root tips in a liquid nutrient solution composed of inorganic salts, sucrose and yeast extract (1934). He showed that yeast could be replaced with the B vitamins, thiamine, pyridoxine and niacin (1937). He wrote *A Handbook of Plant Tissue Culture* (1943), *The Cultivation of Animal and Plant Cells* (1954). In 1963 he published a low salt nutrient culture medium widely used in plant tissue culture.

**wick:** A filter paper, chromatography paper or a strip of some other material used to support plant tissue above a liquid medium and to transport the water and nutrients to the tissue.

**wild type: 1.** The genotype or phenotype of an organism predominating in the control (standard or wild) population, in its natural element. **2.** A specific gene predominant in this population.

**wilt:** The loss of turgidity of plant organs due to inadequate water supply; as leaves and stems may become flaccid (droop) after a drought or due to some types of infection (**wilts**). Usually **wilting** is the result of more transpirational water loss from leaves then water uptake by roots.

**w/m² or watt per square metre:** A common unit of light measurement.

**woody plant medium (WPM):** A modified MS (1962) medium developed for woody plant species by G. Lloyd and B. McCown (1980). It has less nitrogen (both ammonium and nitrate), sodium, potassium and chloride than does MS (1962) medium. It has been increasingly used for the commercial propagation of ornamental trees and shrubs.

**wound tissue:** See **callus**.

**WPM:** See **woody plant medium**.

**w/v:** Weight in volume; as the number of grams of constituent in 100 $cm^3$ (mls) solution.

# X

**xanthophyll:** One of a class of carotene-derived hydrocarbons which function as accessory pigments, or in some algae as primary light absorbing pigments, in photosynthesis.

**xerograft:** See **heterograft**.

**xerophyte:** A plant adapted for dry-climate (**xeric**) conditions, tolerant of drought; as are desert plants like cacti.

**x-ray or Roentgen ray (r):** Electromagnetic radiation of short wave length produced by high speed electrons impacting on a metal object under vacuum.

**xylan:** A polysaccharide of xylose and a component of hemicelluloses.

**xylem:** One of two components of vascular tissue; the woody portion. It is a complex tissue made up of fibres, vessel members, tracheids and parenchyma cells. It conducts water and mineral salts and lends structural support to plant organs. **Protoxylem** is the first (primary xylem) to differentiate and become mature. **Metaxylem** matures later. Secondary xylem is derived from the vascular cambium and occurs in most gymnosperms and dicotyledons. Fibres are elongate cells with pointed ends, thick walls and pits with small borders. Tracheids and vessel elements are tracheary elements having elongated cells with thick lignified walls, and are the principal internal conduits for transpiration. Tracheids are single cells pointed and imperforate at both ends but with walls supplied with bordered pits. Vessels are chains of cells connected through pores or perforations and found only in angiosperms; gymnosperms and pteridophytes have no such vessels.

**xylene or xylol or dimethylbenzene ($C_8H_{10}$ mw 106.17):** A clear liquid solvent used as a clearing agent and paraffin solvent in microscopy.

**xylose ($C_5H_{10}O_5$ mw 150.1):** An aldopentose sugar (wood sugar) present in the woody tissues of many plants as polymeric xylan. It is occasionally used as a carbohydrate source in plant tissue culture media.

# Y

**yeast:** One of a large number of unicellular ascomycete fungi. They are common contaminants of plant tissue cultures.

**yeast extract:** A complex, undefined, B vitamin-containing addendum to some plant tissue culture media.

**yield test:** Productivity assessment.

# Z

**zeatin or 2-methyl-4-(1*H*-purin-6-ylamino)-2-buten-1-ol (Z, $C_{10}H_{13}N_5O$ mw 215.21):** A natural cytokinin first isolated from corn. It is sometimes included in plant tissue culture media. It dissolves in acid (HCl ca. 1 M).

**zeatin riboside (mw 351.4):** A naturally occurring cytokinin sometimes used in plant tissue culture media.

**Zephiran:** The brand name for a disinfecting agent for plant material containing benzalkonium chloride.

**Ziehl's stain:** See **carbolfuchsin**.

**zinc (Zn aw 65.37 an 30):** A hard, blue-white metallic element. It is a microelement essential in chlorophyll; the auxin IAA synthesis; and in several enzyme systems for plant growth. It is included in plant tissue culture media as **zinc chloride** or **zinc sulphate**.

**Zn:** The chemical symbol for the element **zinc**.

**zygote:** A diploid cell arising from gametic fusion; as the fertilized egg, prior to cleavage. It develops into the embryo (**zygotic** embryo).

**zygotic embryo:** See **embryo**.

# Sources


Agrios, G.N. 1964. *Plant Pathology.* Academic Press. U.S.A.

Ambercrombie, M., C.J. Hickman and M.L. Johnson. 1980. (7th ed.). *The Penguin Dictionary of Biology.* Penguin Books Ltd. Middlesex, England.

Ball, E. 1946. Development in sterile culture of stem tip and adjacent regions of *Tropaeolum majus* L. and of *Lupinus majus* L. *Am. J. Bot.* 33: 301–318.

Bergmann, L. 1960. Growth and division of single cells of higher plants in vitro. *J. Gen. Physiol.* 43:841–851.

Bhojwani, S.S. and M.K. Razdan. 1983. *Plant Tissue Culture: Theory and Practice. Developments in Crop Science (5).* Elsvier.

Blackmore, S. and E. Tootill (eds.) 1984. *The Penguin Dictionary of Botany.* Penguin Books Ltd. Middlesex, England.

Brainerd, K.E. and L.H. Fuchigami. 1981. Acclimatization of aseptically cultured apple plants to low relative humidity. *J. Amer. Soc. Hort. Sci.* 106:515–518.

Braun, A.C., P.R. White. 1943. Bacteriological sterility of tissues derived from crown-gall tumors. *Phytopathology* 33:85–100.

Cocking, E.C. 1960. A method for the isolation of plant protoplasts and vacuoles. *Nature.* 187:927–929.

Conger, B.V. 1981. (ed.). *Cloning Agricultural Plants Via In Vitro Techniques.* CRC Press Inc. Boca Raton, Florida, U.S.A.

Cutler, E.G. 1969. *Plant Anatomy: Experiment and Interpretation.* Part 1. Cells and Tissues. Addison-Wesley Co. U.S.A.

De Fossard, R.A. 1976. *Tissue Culture for Plant Propagators.* University of New England, Armidale. N.S.W. Australia.

Dixon, R.A. (ed.) 1985. *Plant Cell Culture: A Practical Approach.* IRL Press Ltd. Oxford, Washington D.C.

Dodds, J.H. (ed.) 1983. *Tissue Culture of Trees.* AVI Publishing Co. Connecticut, U.S.A.

Dodds, J.H. and L.W. Roberts. 1985. *Experiments in Plant Tissue Culture.* (2nd ed.) Cambridge University, U.S.A.

Doliner, L.H. and J.H. Borden. 1984. *Pesterms.* Forest Pest Review Committee of B.C., Canada.

Esau, K. 1960. *The Anatomy of Seed Plants.* John Wiley & Sons. Inc. U.S.A.

Evans, D.A., Sharp, W.R., Ammirato, P.V. and Yamada, Y. (eds.). 1983. *Handbook of Plant Cell Culture* Volume 1. *Techniques for Propagation and Breeding.* Macmillan. New York, N.Y. U.S.A.

Federation of British Plant Pathologists. 1973. A guide to the use of terms in plant pathology. *Phytopath. Pap.* No. 17.

Frobisher, M., R.D. Hinsdill, K.T. Crabtree and C.R. Goodheart. 1974. (9th ed.). *Fundamentals of Microbiology.* W.B. Saunders Co. Pa. U.S.A.

Gamborg, O.L. 1966. Aromatic metabolism in plants. II. Enzymes of the shikimate pathway in suspension cultures of plant cells. *Can. J. Biochem.* 44:791–799.

Gamborg, O.L., Miller, R.A. and Ojima, K. 1968. Nutrient requirements of suspension cultures of soybean root cells. *Exp Cell Res.* 50:151–158.

Gautheret, R.J. 1934. Culture du tissu cambial. *C.R. Acad. Sci. Paris.* 198:2195–2196.

Gautheret, R.J. 1939. Sur la possibilité de réaliser la culture indéfinie des tissus de tubercules de carotte. *C.R. Acad. Sci.* 208:118–120.

George, E.F. and Sherrington, P.D. 1984. *Plant Propagation by Tissue Culture. Handbook and Directory of Commercial Laboratories.* Exegetics Ltd. England.

Grant, J. 1972. 4th ed. *Hackh's Chemical Dictionary.* McGraw-Hill Book Co.

Guha, S. and Maheshwari, S.C. 1966. Cell division and differentiation of embryos in the pollen grains of *Datura* in vitro. *Nature*. 212:97–98.

Hannig, E. 1904. Zur Physiologie pflanzlicher Embryonen. I. Uber die cultur von Cruciferen-Embryonen ausserhalb des Embryosacks. *Bot. Ztg.* 62:45–80.

Harris, D. 1975. *Hydroponics*. Coles Pub. Co. Toronto, Canada.

Hartman, H.T. and D.E. Kester. 1983. *Plant Propagation, Principles and Practices.* (4th ed.). Prentice-Hall, Inc., Englewood Cliffs, New Jersey, U.S.A.

Hawley, G.G. 1981. Revised by. *The Condensed Chemical Dictionary* (10th ed.). Van Nostrand Reinhold Co. Inc. New York, N.Y. U.S.A.

Heller, R. 1953. Recherches sur la nutrition minérale des tissus végétaux cultives in vitro. *Ann. Sci. Natl. Biol. Veg.* 14:1–223.

Jones, Jr., J.B. 1983. *A Guide for the Hydroponic & Soilless Culture Grower.* Timber Press. Portland, Oregon. U.S.A.

Kester, D.E. 1983. The clone in horticulture. *HortScience* 18:831–837.

Knop, W. 1865. Quantitative Untersuchungen über die Ernahrungsprocesse der Pflanzen. *Landwirtsch. Vers. Stn.* 7893 (Cited by Bhojwani, S.S. and M.K. Razdan, 1983.)

Kyte, L. 1983. *Plants From Test Tubes: An Introduction to Micropropagation.* Timber Press. Portland, Oregon. U.S.A.

Lehninger, A.L. 1973. *Short Course in Biochemistry.* Worth Pub. Inc. New York, N.Y. U.S.A.

Little, R.J. and C.E. Jones. 1980. *A Dictionary of Botany.* Van Nostrand Reinhold Co. New York, N.Y. U.S.A.

Lloyd, G. and McCown, B. 1980. Commercially feasible micropropagation of mountain laurel, *Kalmia latifolia*, by use of shoot-tip culture. *Proc Int. Plant Prop. Soc.* 30:421–427.

Martin, E.A. (ed.) 1984. *Dictionary of Life Sciences.* 2nd ed. Pica Press. New York. U.S.A.

Mehra, A. and Mehra, P.N. 1974. Organogenesis and plantlet formation in vitro in almond. *Bot.Gaz.* 135:61–73.

*Merriam-Webster Dictionary (The).* 1974. Pocket Books, New York. U.S.A.

Miller, C.O., Skoog, F., Okumura, F.S., Von Saltza, M.H. and Strong, F.M. 1955. Structure and synthesis of kinetin. *J. Am. Chem. Soc.* 77:2662–2663.

Miller, C.O., Skoog, F., Von Saltza, M.H. and Strong, F.M. 1955. Kinetin, A cell division factor from deoxyribonucleic acid. *J. Am. Chem. Soc.* 77:1392.

Morel, G. 1960. Producing virus-free cymbidium. *Am. Orchid Soc. Bull.* 29:495–497.

Morel, G. and Martin, C. 1952. Guérison de dahlias attients d'une maladie a virus. *C.R. Acad. Sci.* 235:1324–1325.

Murashige, T. and F. Skoog. 1962. A revised medium for rapid growth and bio assays with tobacco tissue cultures. *Physiol. Plant.* 15:473–497.

Muir, W.H. 1953. *Culture conditions favoring the isolation and growth of single cells from higher plants in vitro.* Ph.D. Thesis. Univ. of Wisconsin. U.S.A.

Muir, W.H., Hildebrandt, A.C. and Riker, A.J. 1954. Plant tissue cultures produced from single isolated plant cells. *Science* 119:877–878.

Nitsch, J.P. 1951. Growth and development in vitro of excised ovaries. *Am. J. Bot.* 38:566–577.

Nitsch, J.P. and Nitsch, C. 1969. Haploid plants from pollen grains. *Science* 163:85–87.

Nobécourt, P. 1937. Cultures ensérie de tissus végétaux sur milieu artificiel. *C.R. Séanc. Soc. Biol.* 205:521–523.

Potrykus, I., C.T. Harms and H. Lörz. 1979. Multiple-drop-array (MDA) technique for the large scale testing of culture media variations in hanging microdrop cultures of single cell systems. I: The technique. *Plant Science Letters* 14:231–235.

Raven, P.H., R.F. Evert and H. Curtis. 1976. *Biology of Plants.* (2nd ed.). Worth Publ. Inc. New York, N.Y., U.S.A.

Reinert, J. 1958. Morphogenese und ihre Kontrolle an Gewebekulturen aus Karotten. *Naturwissenschaften* 45:344–345.

Reinert, J. and M.M. Yeoman. 1982. *Plant Cell and Tissue Culture. A Laboratory Manual.* Springer-Verlag. Berlin, Heidelberg, Germany.

Rieger, I., A. Michaelis and M.M. Green. 1976. *A Glossary of Genetics and Cytogenetics* (4th ed.). Springer-Verlag. New York. Inc.

Sharp, W.R., Evans, D.A., Ammirato, P.V. and Yamada, Y. (eds.). 1983. *Handbook of Plant Cell Culture* Volume 2. *Crop Species.* Macmillan. New York, N.Y. U.S.A.

Skoog, F. and Miller, C.O. 1957. Chemical regulation of growth and organ formation in plant tissues cultured in vitro. *Symp. Soc. Exp. Biol.* 11:118–131.

Skoog, F. and Tsui, C. 1948. Chemical control of growth and bud formation in tobacco stem segments and callus cultured in vitro. *Am. J. Bot.* 35:782–787.

Steen, E.B. 1976. *Dictionary of Biology.* Barnes and Noble Books. Harper and Row. U.S.A.

Steward, F.C., Caplin, S.M. and Miller, F.K. 1952. Investigations on growth and metabolism of plant cells. I. New techniques for the investigation, metabolism, nutrition and growth in undifferentiated cells. *Ann Bot.* 16:57–77.

Steward, F.C., Mapes, M.O. and Mears, K. 1958. Growth and organized development of cultured cells. II. Organizations in cultures grown from freely suspended cells. *Am. J. Bot.* 45:705–708.

Sunderland, N. and Roberts, M. 1977. New approach to pollen culture. *Nature* 270:236–238.

Terminology Committee, Tissue Culture Association. 1985. Usage of vertebrate, invertebrate and plant cell, tissue and organ culture terminology. *I.A.P.T.C. Newsletter* 45:15–22. Reprinted from In Vitro 20:19–24 (1984).

Thomas, E. and M.R. Davey. 1975. *From Single Cells to Plants.* Wykeham Publications (London) Ltd.

Thorpe, T.A. (ed.). 1981. *Plant Tissue Culture. Methods and Applications in Agriculture.* Academic Press. U.S.A.

Tomes, D.T., B.E. Ellis, P.M. Harney, K.J. Kasha and R.L. Peterson (eds.). 1982. *Application of Plant Cell and Tissue Culture to Agriculture and Industry.* University of Guelph, Guelph, Ontario, Canada.

Uvarov, E.B., D.R. Chapman and A. Isaacs. 1979. *The Penguin Dictionary of Science.* Penguin Books Ltd. England.

Van Overbeek, J., Conklin, M.E. and Blakeslee, A.F. 1941. Factors in coconut milk essential for growth and development of very young *Datura* embryos. *Science* 94:350–351.

White, P.R. 1934. Potentially unlimited growth of excised tomato root tips in a liquid medium. *Plant Physiol.* 9:585–600.

White, P.R. 1937. Vitamin $B_1$ in the nutrition of excised tomato roots. *Plant Physiol.* 12:803–811.

White, P.R. 1943. *A Handbook of Plant Tissue Culture.* Jaques Cattell Press, Lancaster, P.A. U.S.A.

White, P.R. 1954. *The Cultivation of Animal and Plant Cells.* The Ronald Press. New York, N.Y.

White, P.R. 1963. *The Cultivation of Animal and Plant Cells.* The Ronald Press. New York, N.Y.

Wilson, C.L., W.E. Loomis and T.A. Steeves. 1977. *Botany* (5th ed.). Holt Reinhart and Winston. U.S.A.

Windholz, M., S. Budavari, R.F. Blumetti and E.S. Otterbein (eds.). 1983. *The Merck Index.* Merck and Co. Inc. N.J. U.S.A.